GULOMZhON ShARIFOV
KOMILA NEGMATOVA

MODERN METHODS OF DEVELOPMENT

GULOMZhON ShARIFOV
KOMILA NEGMATOVA

MODERN METHODS OF DEVELOPMENT

OF SALT-RESISTANT COMPOSITE CHEMICAL REAGENTS BASED ON LOCAL AND SECONDARY RAW MATERIALS

ScienciaScripts

Imprint
Any brand names and product names mentioned in this book are subject to trademark, brand or patent protection and are trademarks or registered trademarks of their respective holders. The use of brand names, product names, common names, trade names, product descriptions etc. even without a particular marking in this work is in no way to be construed to mean that such names may be regarded as unrestricted in respect of trademark and brand protection legislation and could thus be used by anyone.

Cover image: www.ingimage.com

This book is a translation from the original published under ISBN 978-620-5-51380-4.

Publisher:
Sciencia Scripts
is a trademark of
Dodo Books Indian Ocean Ltd. and OmniScriptum S.R.L Publishing group
Str. Armeneasca 28/1, office 1, Chisinau MD-2012, Republic of Moldova, Europe
Printed at: see last page
ISBN: 978-620-5-39664-3

MINISTRY OF HIGHER EDUCATION OF THE REPUBLIC OF UZBEKISTAN

JIZZAKH STATE PEDAGOGICAL UNIVERSITY

ABDULLAH KADIRI INSTITUTE

SHARIFOV GULOMJON NABIEVICH
NEGMATOVA KOMILA SOYIBZHANOVNA

MODERN METHODS FOR DEVELOPING SALT-RESISTANT COMPOSITE CHEMICAL REAGENTS BASED ON LOCAL AND SECONDARY RAW MATERIALS

UDC 678.743

Sharifov Gulomjon Nabievich - Doctor of Philosophy (PhD) in Chemistry, Senior Lecturer of the Department of Chemistry and methods of its teaching at Jizzakh State Pedagogical University, Jizzakh

Negmatova Komila Soyibzhanovna - Doctor of Technical Sciences, Professor, Head of the Laboratory of Composite Materials of the State Unitary Enterprise "Fan va Tarakkiyot at the I. Karimov Tashkent State Technical University, Tashkent

THE monograph describes modern methods of research of complex regularities of physical-mechanical and operational properties change as well as dependence of their properties on type, composition and content of ingredients in order to develop optimal compositions of rapastable composite chemical reagents and drilling fluids based on them and technological processes for application in the process of oil and gas wells drilling under rapture conditions. Also, modern methods of physical-chemical, technological analysis, infrared spectroscopy, X-ray phase, differential-thermal analysis and other standard methods of analysis for the development of efficient import-substituting, non-deficit, accessible, relatively cheap and environmentally friendly chemical reagents for stabilization of drilling mud on the basis of available raw materials have been used.

For specialists in technological chemistry, ecology, engineers, technicians and laboratory technicians working in the field of petrochemistry, doctoral and undergraduate students in the same fields.

Responsible editor:
D. in Chemical Sciences ***N. I. Muminova***

Reviewers:
Doctor of Chemical Sciences ***M.M. Sultonov***
Doctor of Technical Sciences ***S. S. Negmatov***

Monograph recommended for publication by the Scientific Council of the State Unitary Enterprise "Fan va tarakkiyot" at the Tashkent State Technical University (Minutes No. 3 of 10.10.2022)

DEFINITION

I N T E R

At present in the world there is a growing demand for rapastable chemical reagents and drilling agents on their basis used for oil and gas wells drilling in conditions of rapaciousness on areas of saline sediments, consisting mainly of different salts melts. Therefore, for drilling of saline oil and gas wells the creation of efficient rapastable composite chemicals and drilling muds on their basis, which dissolve and neutralize magnesium chloride salt solution ($MgCl_2$), magnesium sulphate ($MgSO_4$) and other metal salts, leading to the increase of rock destruction tool speed and efficiency of oil and gas wells drilling process is of special importance.

Worldwide research is being carried out to develop chemicals and drilling fluids used in oil and gas wells in saline areas, to ensure stability of drilling fluids, increase effectiveness of rock destruction tools, prevent shrinkage and provide stability of wellbore walls and cuttings out of wells. In this aspect, special attention is paid to creation of effective rapastable chemical reagents, capable of replacing imported expensive chemical reagents used in drilling fluids, in order to increase mechanical speed of drilling rigs and stability of drilling fluids of oil and gas wells.

A number of activities are being carried out in the Republic to develop rapastable composite chemical reagents based on local and secondary raw materials, and certain results have been achieved. In the fourth point of the fourth direction of the Strategic Action Programme for the Further Development of the Republic of Uzbekistan on "...effective mechanisms for stimulating research and innovation activities, application of scientific and innovative developments..."[1] sets out important tasks. In this aspect, the development of effective compositions of rapastable composite chemical

[1] Presidential Decree No. UP-4947 on the Action Strategy for the Five Priority Development Areas of the Republic of Uzbekistan for 2017-2021

reagents based on local and secondary raw materials with improved physical-chemical and operational properties, and the technology of making drilling muds on their basis for oil and gas wells drilling in the JSC "Uzbekneftegaz" system is of particular importance.

This monograph, to a certain extent, serves to fulfil the tasks set out in Presidential Decree No. UP-4947 of 7 February 2017 "On the Action Strategy for the Five Priority Development Areas of the Republic of Uzbekistan for 2017-2021", and Cabinet of Ministers Decree No. PP-3983 of 25 October 2018 "On measures to accelerate the development of the chemical industry in Uzbekistan", No. PP-4426 of 24 August 2019 "On further enhancing the responsibility of state, economic management and local executives

In the field of development and creation of composite polymeric chemical reagents significant contributions were made by such scientists as N.S. Enikolorov, A.A. Berlin, V.V. Korshak, E.F. Oleynik, A.F. Matthews, A.I. Bulatov, M.A. Askarov, S.Sh. Rashidova, S.S. Khamraev, A. Akramkhojaev, K.S. Negmatova, V.P. Guro, V.I. Suri, M.M. Sharma, Wawrzos Frank A. and others.

The works of S.S. Negmatov, G. Rakhmanberdiyev, A.H. Yusupbekov, A. Agzamkhodjaev, A. Tokunov, Y.M. Basarigin, R.Z, Sharafutdinov, A.A. Askadski, H.C. Darley, M.T. Chapman, H.C. Darley, N.H. Growcock and many others.

Taking into account the analysis of existing works, it is necessary to note, that for wide application of highly stable drilling muds, including rapastable ones, it is necessary to deepen research in the sphere of modification of reagents, creation of compositions, allowing to regulate and improve technological properties, used for rapastable drilling agents according to requirements, occurring under difficult mining-geological conditions of drilling. The analysis showed that there is almost no data concerning influence of components on their rapastable properties and scientifically substantiated approach for controlling physical-chemical and technological parameters of rapastable drilling muds on the basis of rapastable composite chemical reagents. The task of developing effective compositions of composite Rapastable chemical reagents on the basis of local raw materials and production wastes and getting high-quality drilling agents, allowing to conduct drilling process without complications and accidents, is far from being solved. The present work is devoted to solving these problems.

The subject of the research is the study of complex regularities of physical-mechanical and performance properties change as well as dependence of their properties on type, composition and content of the ingredients with the purpose of development of optimal compositions of rapastable composite chemical reagents and drilling muds on their basis and technological processes for application in drilling oil and gas wells under rapture conditions.

In addition, modern methods of physical-chemical, process analysis, infrared spectroscopy, X-ray diffraction, differential-thermal analysis and other standard methods of analysis have been used.

The scientific novelty of the study lies in the following:

New science-based approach to development of rapastable composite chemical reagents based on establishing main regularities of interaction of composition components as well as influence of nature, type, structure, content of organic-mineral ingredients and their ratios on physical-chemical and technological properties of composite reagents has been developed;

The main characteristics of changing physical-chemical and technological parameters of rapastable composite chemical reagents and drilling muds based on them, such as density, viscosity, water yield, static shear stress, depending on the structure, composition and correlation between the components of the composition have been revealed;

For the first time new compositions of import-substituting rapa-resistant composite chemical reagents based on organomineral ingredients from local raw materials and production wastes with high water-solubility, thermodynamic stability, availability, sufficiently low cost and preparation of drilling muds on their basis, capable of working with mineralized formation water in ravine conditions are developed;

scientific-methodological principles of production technology of rapa-resistant composite chemical reagents and drilling muds based on them, capable of operating with saline formation water in conditional rapa-production, as well as being economical and ecologically safe, have been developed;

it has been established that the developed rapastable composite chemical reagents and drilling fluids based on them have shown high stabilizing properties in the process of oil and gas wells drilling in the republic due to their good structural-chemical features.

The practical results of the study are as follows:

On the basis of the research results, import-substituting rapastable composite chemical reagents and drilling fluids used in oil and gas wells drilling have been developed by purposeful selection of type, content and ratio of organic-mineral ingredients based on local raw materials and production wastes;

Rapa-resistant composite chemical reagents and drilling muds based on them have been developed that are capable of exerting a strong stabilizing effect on physical-chemical and technological properties of drilling fluids due to the synergistic effect during drilling of oil and gas wells in conditions of rapa-permeability.

Reliability of the obtained results is based on the results of several laboratory and industrial experiments carried out by the author using modern methods of physical and chemical analysis in the complex processing of local raw materials and industrial wastes and using modern computers and software, preparation of composite chemicals for oil and gas industry.

The scientific significance lies in the fact that by determining the regularities of the influence of the type, structure, chemical nature, composition, ratio of the ingredients and establish the basic physical, chemical and technological properties, as well as define the scientific basis for the creation of rapastable composite chemical reagents that combine mineral and organic ingredients in their composition.

The practical value of the work lies in the use of created rapastable composite chemical reagents of KCR-RUS class, which increase the stability of solutions during drilling wells and the mechanical speed of the drilling process by 10-15 %, increase the opening of productive horizons for oil and gas by 30-35 %, and also provide economic efficiency of ecological safety of the environment.

The following results were obtained on the basis of scientific research on the development of rapastable composite chemical reagents based on local and secondary raw materials:

Patent for invention of Intellectual Property Agency of the Republic of Uzbekistan №IAP 050462015 "Method of preparation of powdery water-soluble gossypol resin" has been granted. The results allowed to develop method of effective composition of rapastable composite chemical reagents based on local raw materials and industrial waste with high physico-chemical and technological properties;

Resource-saving technology of rapastable composite chemicals is introduced into drilling of oil and gas wells in Uzbekneftegaz JSC (reference of Uzbekneftegaz JSC 03-17-5/165 dated October 27, 2021). As a result, more effective compositions of rapastable composite chemical reagents in the process of drilling oil and gas wells were obtained;

The developed rapastable composite chemical reagents based on local raw material and industrial wastes with high physical-chemical and technological properties are introduced in Uzbekneftegaz system (reference of Uzbekneftegaz 03-17-5/165 dated 27 October 2021). As a result, contributed to increase of economic efficiency of JSC "Uzbekneftegaz".

CHAPTER I. STATE AND PECULIARITIES OF MINING-GEOLOGICAL CONDITIONS OF WELLS DRILLING IN SALINE AREAS OF THE REPUBLIC AND JUSTIFICATION OF DEVELOPMENT OF EFFECTIVE COMPOSITE CHEMICAL REAGENTS FOR DRILLING MUD STABILIZATION

§1.1 A brief geological description of the saline sediments of the oil and gas bearing areas

Most of the oil and gas fields in Southwest Uzbekistan are confined to Upper Jurassic sediments, which are overlain by thick chemogenic deposits of Kimmeridge-Titon in the form of gypsum, anhydrite and rock salt (Gaurdak Formation). The upper part of the formation is mainly composed of pink coarse-crystalline layered rock salt with interlayers of anhydrite, potassium and magnesium salts (carnallite and bischofite) with interlayers of red clay. The thickness of these deposits ranges from 200 to 600 meters. The Gaurdak Formation sediments often contain lenses of brine with pressures that, in some cases, are as high as full mountain [2; c 5].

Analysis of field materials has shown that chemogenic rocks constitute a significant part of the geological section of many oil and gas bearing areas in the Beshkent trough. They occur in the form of saline domes, interlayers of rock salt, gypsum and anhydrite, less frequently included in clay-sandy rocks. It is well known that chemogenic deposits are good covers for accumulation of hydrocarbons, so the volume of drilling for subsalt deposits is increasing. In Uzbekistan, chemogenic zones are confined to Upper Jurassic (Kimerij-Titan) deposits in the Amu Darya and Afghan-Tajik Depression and Neogene deposits in the Fergana Depression [3; p. 688, 4; p. 272].

In the Amudarya depression, chemogenic rocks lie in the form of bundles at depths of 1000-3000 m, 50 to 1000 m thick, with alternating interlayers of chlorides and sulphates. In Chardzhou stage and Beshkent trough it is common to distinguish upper and lower layers of rock salt separated by gypsum-anhydrite layer. In some areas (Pamuk, Urta-Bulak, Zaverdy, Kultak, etc.), brine brine lenses are often uncovered in this sedimentary sequence. The brine brine lenses are confined to isolated local areas in the salt column and have limited dimensions and pressure close to geostatic pressure. By composition the brine brines are chloride-sodium-calcium and chloride-sodium-magnesium salts with mineralization of 300-670 g/l and density at 20°C of 1.28-1.38 g/cm3^3 . In the interval of rock salt occurrence the bottom-hole temperature is 120-140°C [5; p.60, 6; p.107].

In the Fergana Depression, salts occur in the Neogene at depths of 2000 to 5500 m (Chust-Pap, Mingbulak Pl.). Deposits are mainly sandy-clayey rocks oversaturated with salt, thickness of deposits is up to 2000 m and more, bottom-hole temperature is 180°C. Brine beds with high salinity and density up to 1.27 g/smq^3 have been uncovered in these sediments in the Chust-Pap area. The anomalous reservoir pressure coefficient has been observed to reach 1.9.

Drilling of rock salt (halite) and anhydrite deposits is in turn complicated by interlayers and lenses of carnallite, bischofite and minerals of complex composition, as well as their brines. In saline formations the further deepening of the borehole is often terminated because of complications arising in the borehole; also strings crushing may be observed (Zaverdy and Kultak) [7; p. 632, 8; p. 190].

The technology of drilling wells in saline sediments is one of the difficult and less solved issues of drilling technology. Therefore, when choosing rational methods of drilling out saline sediments it is necessary to consider their physical and chemical properties determined by the mineralogical composition of salts, conditions of occurrence (depth, temperature, pressure), as well as the degree of contamination by various impurities. Borehole drilling in chemogenic deposits is connected with overcoming of some negative phenomena: negative influence of salts on properties of drilling mud; formation of caverns and constrictions in the borehole connected with dissolution and plastic flow of salts, especially such highly soluble and highly plastic ones as potassium-magnesium [11; p. 35-40, 10; p. 28].

§ 1.2 Composition of chemogenic rocks

Chlorides and sulphates of sodium, potassium, magnesium and calcium are the main rock-forming minerals. Chlorides include halite (NaCl), sylvinite (KC1), sylvinite (a mixture of sylvin dominated by halite), carnallite (KCl-$MgCl_2$ -$6H_2O$), bischofite ($MgCl_2$ -$6H_2O$), sulfate-angidrite ($CaSO_4$), gypsum ($CaSO_4$ -$2H_2O$), kieserite ($MgSO_4$ -H_2O), langbeinite (K_2SO_4 -$2MgSO_4$), polyholite (K_2SO_4 $MgSO_4$ $CaSO_4$ -$2H_2O$), kyanite (KCl-$MgSO_4$ -$3H_2O$), etc. In chemogenic sediments other minerals (quartz, carbonates, clay minerals) are often included as admixtures and separate inclusions [12; p. 99-102, 13; p. 201-202].

There are several types of halide-bearing rocks:

1. Rock salt. The drilling mud must be salted in order to prevent salt dissolution and cavern formation, which can in turn lead to scaling and collapse of the overlying terrigenous rock. Drilling can be carried out with brine washing if the required density does not exceed 1,2 g/cm^3 [14; p. 355-356, 15; p. 250-252];

2. Rock salt with interlayers of bischofite and other salts. These rocks must be drilled using brine or clay mud containing salt with greater solubility;

3. Rock salt with interbedded terrigenous rocks. These require the use of salt-saturated drilling muds, which are chemically treated to produce low water yields

4. Rock salt with interlayers of bischofite and terrigenous rocks. They must be drilled with flushing with chemically treated solutions with low water yield, salinated with magnesium chloride see[3] [14; p. 355-356, 15; p. 250-252];

When saline deposits are penetrated with water-based drilling fluids, conditions are created for dissolution of these rocks, resulting in a widening of the borehole and formation of caverns, which slows the rate of incoming mud flow, worsens conditions for removal of drilled-out rocks, which leads to formation of slurry plugs. The solubility of salts depends on their chemical composition and temperature. According to solubility in water they can be arranged in the following order: bischofte, kieserite, sylvin, halite, anhydrite, gypsum [17; p.18, 18; p.151-152].

The solubility of most salts increases with increasing temperature, as shown in Table 1.1.

Some salts, such as kieserite, are characterised by a maximum solubility at elevated temperatures. Rocks such as sylvinite and carnallite dissolve differently. In the case of sylvinite the solubility of KCl in water increases significantly with increasing temperature, while the solubility of NaCl changes little. At saturation of KCl solubility of NaCl decreases with heating. Thus, a solution saturated with both salts at 20 °C contains 10.4 % ÊÑl and 20.7 % NaCl, and at 100 °C - 21.7 % ÊÑl and 16.8 % NaCl. When carnallite interacts with water, $MgCl_2$ enters the solution, while KCl remains in the precipitate.

The difference in solubility of salts in water and the effect of temperature on this property results in a complex change in the salt composition of the washing fluid as the well goes deeper and the downhole temperature rises, resulting in a change in the salt composition of the solution along the wellbore.

Table 1.1

Solubility of salts depending on temperature (g per 100 g of solution)

Temperature, °C	Name of salts						
	halite $NaCl$	sylvin KCI	bischofite $MgCl_2$ -6H_7 0	gypsum $SaSO_4$ 2H_2 0	anhydrite $SaSO_4$	kieserite $MgSO_4$ H_2 O	gypsum $SaSO_4$ 0.5H_2 O
0	23,3	21,9	34,5	0,1759	-	18	-
10	26,3	23,8	34,9	0,1928	-	22	-
18	-	-	-	0,2016	-	-	-
20	26,4	25,5	35,3	-	0,298	25,2	-
22	-	-	35,6	-	-	-	-
25	-	-	36,2	0,208	0,274	26,7	-
30	26,5	27,1	35,8	0,209	-	28,0	-
35	-	-	-	0,2096	0,242	29,3	-
40	26,7	28,6	36,5	0,2097	-	20,8	-
45	-	-	-	-	0,201	29,83	-
50	26,8	30,0	37,2	-	-	-	-
60	27,0	31,4	37,9	-	-	33,42	-
75	-	-	-	0,1847	0,144	37,42	-
80	27,5	33,9	39,8	-	-	-	-
90	-	-	41,0	-	-	34,6	-
100	28,1	36,0	42,2	0,1619	0,67	33,5	0,1645
120	-	-	-	-	0,0435	30,0	0,103
130	-	-	48,58	-	0,035	-	0,083
140	-	-	-	-	0,028	24	0,0665
150	-	-	51,81	-	0,0222	-	0,053
170	-	-	54,55	-	0,014	8	0,325
180	-	-	-	-	0,1112	5	0,0205
190	-	-	-	-	0,0092	2,5	0,0165
200	-	-	56,8	-	0,0076	1,5	-
220	-	-	59,51	-	0,0076	0,8	-
240	-	-	-	-	0,0055	0,5	-
244	-	-	-	0,0073	-	-	-
300	-	-	67,84	-	-	-	-
316	-	-	-	0,0042	-	-	-

The temperature and solubility of the mud change as it circulates. Therefore, selective dissolution or crystallisation of salts is continually occurring along the borehole. For each temperature, there is a well-defined relationship between the concentrations of each of the dissolved salts. The excess or lack of any salt in the solution leads to changes in concentration of the other salts [19; p. 200-202; 20; p. 105-107].

The high bottomhole temperature and the presence of polymineral salts make well drilling in chemogenic deposits even more difficult. All this leads to the necessity of frequent treatment of the solution, systematic work-overs of the wellbore, i.e. long and expensive unproductive works and often ends with wells abandonment [21; p. 302-303].

A decrease in the temperature of the incoming flushing fluid flow causes crystallization of salts, mainly sodium chloride, and their accumulation as sludge on the borehole walls, reducing its diameter, as well as in the surface circulation system. At low flow rates, sludge plugs from crystallised salt can occur at the top of the borehole.

The experience of drilling deep wells in sulphate-halide deposits shows, that NaCl concentration in the solution saturated with this salt increases up to 35-36 % in the process of drilling. While drilling into the whole thickness of chemogenic deposits the concentration of ÊÑl in the solution changes from 0 to 28 %, and $MgCl_2$ - ÊÑ1 - up to 14 %. An increase in the concentration of these salts is accompanied by a corresponding decrease in the concentration of rock salt - from full saturation to 18 %, and in some cases to 7 %, which is a consequence of the influence of temperature on solubility. It should be taken into account that insignificant under-saturation of NaCl solution, especially under high temperature conditions, leads to appreciable increase of solubility of ÊÑl, $MgCl_2$, Na_2SO_4 , $MgSO_4$ and, consequently, promotes selective cavern formation in formations, containing these salts, despite the fact that flushing fluid pumped into the well is enriched with rock salt to saturation. This effect is more pronounced the greater the difference in temperature at the bottomhole and at the wellhead. Bischofite and carnallite are more sensitive to unsaturation of the drilling fluid than other salts. For dissolving 100 g of bischofite, for example, at temperatures 0; 50 and 100°C it is necessary respectively 35,8; 26,4 and 0 cm of water[3] [22; p.90-101].

Thus, dependence of salt solubility value on temperature causes cavern formation in deep chemogenic formations, on the one hand, and wellbore narrowing in the upper part of the wellbore, on the other hand. To prevent complications when drilling wells in combined chemogenic deposits, related to salt solubility, it is necessary to select such salt composition of flushing fluid, which best corresponds to the nature of drilled rocks [23; p. 120-122].

In addition, we analyzed the Aizavod area, which is located within the Chorjau stage of the Bukhara-Khiva oil and gas bearing province, in the central part of the Beshkent trough within the Shakarbulak rampart. Regionally productive in this area are subsalt carbonate deposits of Upper Jurassic Kellovay-Oxfordian stage. To date, a number of deposits have been identified within the Beshkent Trough:

- Gas condensate fields: Beshkent, Shurtan, Buzahr, north Guzar, etc. Guzar, etc;

- Oil and gas condensate fields: Shakarbulak, North Shurtan and Mangit;

- oil: Garmiston and Kumuk.

However, due to geological conditions, the following complications may occur during exploration drilling at the Okaltyn, Central Yangikent, Mangit, Aizovat 1p and other areas:

1. Drilling of Neogene-Quaternary and Eocene-Oligocene polygenic deposits may result in wellhead erosion during drilling, rock slides and collapses of the borehole walls, and partial absorption of flushing fluid.

2. When uncovering Bukhara Pamocene limestones with low-viscosity drilling fluids with specific gravity of 1.16-1.18 g/cm^3 , flushing fluid may be absorbed, and in case of lower specific gravity, water leakage may occur.

3. Drilling of boreholes in Cretaceous sediments proceeds normally without any particular complications, but in clays of Alpine and Turonian stages with clay mud density below 1.20-1.24 g/cm^3 collapse of borehole walls, cavernous and trough formation and sticking of drilling tools are possible.

4. Drilling in the Senomo, Alpine and Neocomian Nadjarous sandstones may result in borehole constrictions, partial fluid intakes and water perforations.

5. When drilling through salt-and-anhydrite deposits of cimmeridge-titanium strata, a sharp deterioration of flushing fluid quality, formation of caverns and troughs, rasping, coagulation of clay mud and wellbore narrowing in intervals of anhydrite interlayers, partial absorption of drilling mud are possible.

6. When penetrating and penetrating of Upper Jurassic deposits of Kelloway-Oxford stage, irrespective of the type of opened carbonate section, in case of deviation of drilling mud parameters from the design parameters oil-gas-emitting, absorption of clay mud with intensive crusting on the borehole walls and sticking of drilling tools are possible [24; p. 54, 25; p. 105-107].

It should be noted that the density and formulation of the drilling fluid may change during the deepening of the well, depending on the complications encountered.

The upper anhydrites are dense, firm anhydrites, grey, brown with frequent interlayers of dark brown and red-brown clays. The thickness of the salt and anhydrite formation at the Aizovat Pl. is assumed to be 500-550 m. This can be seen from Table 1.3. Tectonically, Aizovat area is located in the central part of Beshkent megaproject, which occupies southeastern part of Chardzhou stage [24; p. 54, 25; p. 105-107].

§ 1.3 State of technology for drilling exploration wells and justification of development of effective chemical reagents for flushing fluids operated during drilling in salt sediments in the conditions of rapa-proliferation

The formulation of drilling fluids for drilling wells in the areas of western Uzbekistan in most cases is very simple and includes bentonite, a double-acting polymer or two polymers, one of which regulates rheological properties and filtration, while the second is a selective flocculant. The protection of the solution against electrolyte and temperature aggression is achieved by treating it with reagents such as starch, CMC-500 and 600 and their combination with KSSB and chrome lignosulphonates [26; p. 82-83].

The task of regulating the structural and mechanical properties of salt-saturated solutions is solved, in addition to the use of Poligorskite, by preliminary dispersion and hydration of clay in fresh water and its stabilization with protective reagents, as well as strengthening the structure formation by emulsifying oil, additives SMAD and humates. In addition the solutions based on hydrogels of salts and, in particular, magnesium hydrogel, obtained by controlled condensation, are used [27; p. 151-152, 28; p. 22].

Solutions with low content of the solid phase, in which up to 3 % of sodium bentonite and no more than 3 % of drilled (other) clay are known. The double-acting polymer is an acrylate and acrylamide copolymer, the amount of which is in the range of 0.01-0.05 % [29; p. 100-1101, 30; p. 48-49].

The use of non-dispersed solutions with low solids content is difficult when drilling anhydrites, especially if there is a need to reduce filtration. The addition of anhydrous soda ash, barium carbonate and water is recommended for low anhydrite capacities. If the calcium content is higher than 0.02 %, a mixture of sodium polyacrylate and CMC is recommended. The amount of dispersant must be kept to a minimum in order to regulate the solids content effectively. Recently, other chemical reagents have been recommended as a component of solutions with low solids content as they are more resistant to calcium salts [31; p. 80].

The type and parameters of the flushing fluid were selected taking into account the expected reservoir pressures in the section, the geological and technical conditions of the project section penetration and also based on drilling experience in neighbouring areas and fields. The parameters of the flushing fluid are given in Table 1.2.

As it is seen from Table 1.2, when drilling under conductor to the depth of 0-150 m the polymer drilling muds with density 1,10-1,12 g/cm^3 ; water productivity 12-14 cm^3 /30 min; nominal viscosity 45-50 sec; pH 8-9 are used. When drilling in Bukhara limestones of Pamocene 150-1300 m deep they use less viscous polymer-humite drilling muds with the following parameters: density - 1,16-1,18 g/cm^3 ; nominal viscosity - 30-50 sec; water-output - 11-12 cm^3 /30 min; pH = 8-10.

Table 1.2

Composition and parameters of washing liquids

Intervals	Flushing Liquids	Flushing liquid parameters				Name of chemicals
		ρ, g/cm^3	T_{500}, C	B, cm^3 /30 min	pH	
0-150	clay mortar	1,1-1,12	45-50	12-14	8-9	caustic soda, soda ash, K-4, K-9, CMC
150-1300	clay mortar	1,16-1,18	30-50	11-12	8-10	soda ash, K-4 or K-9, CMC, caustic soda, lignite
1300-2750	clay solution	1,20-1,24	40-60	8-10	9-10	Na_2CO_3, K-4, K-9, KMC-600, USER, oil
2750-3340	clay solution	1,32-1,34	45-80	8-10	9-10	Na_2CO_3, K-4, K-9, KMC-600, UCP, NaOH, NaCl, starch, oil
3340-4000	clay solution	1,22 - 1,24	40-60	6-8	8-9	Na_2CO_3, K-4, K-9, KMC-600, NaOH. FHLS, oil, chrompea, hypane (before column descent)

Drilling wells in the 1300-2750 m interval in the Alpine and Turanian clays is often accompanied by collapses and ridge and trough formations, which leads to stuck drilling tools.

At that a humate-polymer oil-emulsion mud with density 1.20-1.24 g/cm^3, water yield 8-10 cm^3 /30 min; conditional viscosity 40-60 sec; pH = 9-10 is used. During drilling in 2750-3340 m interval of Neocomian substage and salt-and-anhydrite formation there is observed a sharp deterioration of flushing fluid, which was treated using large amount of starch, oil and polymeric reagents with the following technological parameters: γ - 1,32-1,34 g/cm^3 ; T_{500} - 45-80 sec; B - 8-10 cm^3 /30 min; pH = 9-10. In case of rape the density of the drilling fluid increases from 1,3 to 2,10 g/cm^3 , for stabilization PAA and starch are used, for liquefaction - the chemical agent PCLS [32; p. 2, 33; p. 98-99].

During further deepening of the well in the interval 3340-4000 m, after some time during the drilling process, deterioration of the clay mud quality was noticed, e.g. conditional viscosity increased to non-fluid, water yield to 30 cm^3 /30 min, pH of the mud decreased to 7,5-8. At the same time the clay solution is re-treated with caustic soda and polymer reagent in the presence of FHLS and oil. After that when technological parameters of the mud were achieved with ρ - 1,2-1,24 g/cm^3 ; T_{500} - 40-60 sec; B - 6-8 cm^3 /30 min; pH = 8-9 drilling was continued.

Prospects of oil and gas content in many areas are associated with depths of 5-6 thousand metres. A considerable part of promising horizons is confined to subsalt deposits. That is the reason the protection of drilling agents against aggression of polymineral composition electrolytes at temperature 220-240^0 is of topical importance. Protection of the mud against polymineral aggression is necessary not only at drilling out salt strata, but also when using hard highly mineralized waters for preparing muds in waterless areas [34; p. 165].

To avoid cavern formation, salts are drilled using salt-saturated muds. Depending on reservoir pressures, thickness and composition, saline rocks are drilled with brine, clayey salt-saturated mud not treated with reagents to reduce filtration, and salt-saturated clayey mud stabilised with reagents. These reagent-based formulations are only suitable for low concentrations of calcium and magnesium salts. Attempts have been made to increase the thermal and salt resistance of these reagents by introducing various initiators (reagents carbophene, carbonyl carbodinol) into their structure, but the products obtained do not meet the harsh conditions of mineral aggression [35; p. 74-75].

Thus, the development of new effective heat-soluble composite chemical reagents on the basis of local raw materials and waste products of organic and inorganic origin, used for preparing flushing solutions for drilling oil and gas wells, operated in conditions of depression, is a very urgent problem for the oil and gas industry of the republic.

In this connection this project considers the development of composite chemical reagents based on local raw materials and waste products of organic and inorganic origin used for preparing flushing solutions for drilling oil and gas wells operated in conditions of oil rasping.

§1.4 Conclusions of the first chapter

Based on the analysis of existing works, it is necessary to note, that for wide application of highly stable drilling muds, including rapastable, it is necessary to deepen research in the field of modification of reagents, creation of compositions, allowing to regulate and improve technological properties of used rapastable drilling fluids according to requirements, arising under difficult mining-geological drilling conditions.

CHAPTER II. SELECTION AND JUSTIFICATION OF OBJECTS AND METHODS OF RESEARCH OBJECTS OF RESEARCH AND METHODS OF OBTAINING SAMPLES FOR STUDYING PHYSICO-CHEMICAL PROPERTIES OF CHEMICAL REAGENTS

§2.1 Selection and justification of research objects based on local raw materials and waste products for the development of composite chemical reagents and their characteristics

The study of literature data showed that for oil and gas well drilling processes it is necessary to use a variety of chemicals: CMC, K-4, K-9, HIPAN, PRS, KSSB, NaOH, $CaCO_3$, USHR, sulfanol, graphite and others. Most of them are expensive and many of them are not produced in Uzbekistan but are imported from abroad for foreign currency. It should be noted that both imported chemical reagents complying with European standards and domestic reagents are insufficiently effective, especially when using saline formation water as flushing fluids for drilling and opening productive horizons, in particular, in the Plato-Ustyurt regions of Karakalpakstan and Kashkadarya oblast.

Naturally, there is an urgent need to develop effective composite materials - chemicals that provide the necessary density, low water retention, SNS and other parameters to meet the requirements for drilling fluids.

On the basis of the analysis of works carried out on the use of drilling agents when drilling oil and gas wells it has been established, that the main factors allowing to carry out qualitative complex of drilling works are the following [36; p. 37-39].

Taking into consideration the results of studies published in literature on laboratory, pilot and industrial tests of bit performance while drilling oil and gas wells and their analysis, we have developed the following requirements for the composite chemical reagents and drilling muds based on them, which are shown in Figure 2.1.

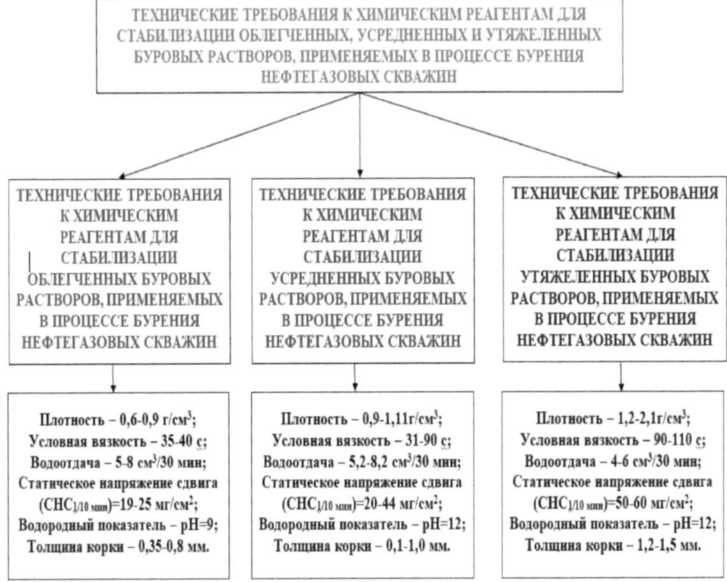

Figure 2.1: Technical requirements for chemical agents for stabilisation of light, medium and weighted drilling fluids used in oil and gas drilling operations

Density - plays a major role in ensuring the smoothness and stability of the oil and gas drilling process. It should be pointed out that the density of the drilling fluid must match the density and pressure of the formation water. The hydrostatic pressure of the drilling fluid must be greater than or equal to the reservoir pressure in order for the bit to operate smoothly and steadily:

$$P_{д.пл.} \leq P_{d.ras.} \quad (1)$$

The hydrostatic pressure of the drilling fluid is determined using the following formula:

$$Pd_{.ras.} = \gamma_{раст.} - g_{расm.} -h = kg/m^3 -m/sec^2 - m = kg/m\text{-}sec^2 = Pa, \quad (2)$$

where $\boldsymbol{P}_{\partial.пл.}$ *is* reservoir pressure, *Pa;*

$\boldsymbol{Pd}_{.ras.}$ - drilling fluid pressure, *Pa;*

$\gamma_{раст.}$ - Mud density, *kg/m^3 ;*

$\boldsymbol{g}_{pacm.}$ - free fall acceleration of drilling fluid, *m/s^2 ;*

h *is* the depth of the working seam, *m.*

The water yield (B, cm^3 /30 min) should be between 2-3, allowed up to 10 cm^3 /30 min. An increase in the water yield of more than 10 cm^3 /30 min increases the hydrophilicity of the mortar. Mud adsorption to the borehole walls occurs and drilling fluid consumption increases, which leads to higher drilling costs for oil and gas wells.

Viscosity (T_{500}) is important for the lifting power of the drilling fluid. The viscosity is determined by the concentration, quality and degree of hydration of suspended mud particles. The viscosity is classified into: *the conventional viscosity* - characterizing the hydraulic resistance of the drilling fluid to the flow; *the effective viscosity* - indirectly characterizing the viscosity of the drilling fluid as a Newtonian fluid; and *the plastic viscosity* - that part of the resistance to the fluid flow caused by mechanical friction [37; p. 44].

The average viscosity of drilling fluids should have values between 36 and 54 s. Light muds should have viscosities between 35 and 45 seconds, normal muds between 45 and 65 seconds and weighted muds between 80 and 110 seconds.

The static shear stress (SST) is the force at which the drilling fluid structure begins to break down, referred to a unit area. The static shear stress is usually expressed in dPa.

The static shear stress determines the ability to hold cuttings and weighting agent particles in suspension during mud stoppages and ensures the stability of the colloidal mud systems. To ensure this capability, the static shear stress must be greater than the force created by the weight of the cuttings or weighting agent particles. Otherwise, these particles, in the absence of drilling fluid circulation, will be deposited in the bottomhole part, which may eventually lead to stuck drillstring with cuttings [38; p. 131-132, 39; p. 28-34].

At the same time, it should be pointed out that with increase of static shear stress, conditions of self-cleaning of drilling mud from cuttings at the surface deteriorate, as well as value of bottomhole pressure pulses increases. At initiation of the drilling fluid flow (when starting up the pump) and during ROP, which, in its turn, increases a probability of fluid eruptions, stability of the borehole walls, hydraulic fractures and drilling fluid absorptions are violated [40; p. 74-75, 41; p. 256-258].

The static shear stress (SST), depending on the concentration of chemical agents in both fresh and saline water-based drilling fluids, should be in the range Θ_1 = 15 to 25 mg/cm^2 , Θ_{10} = Θ_1 +(40-60)% Θ_{1M} . So, the SNS should be minimal, but sufficient to keep the drilled rock and weighting agent particles suspended in the resting drilling fluid [42; p. 300].

A common measure of the electrochemical properties of drilling fluids is **the hydrogen index (pH).** It characterises the concentration of hydrogen ions [H+] in the drilling fluid (degree of acidity or alkalinity of water-based drilling fluids): pH = 7 - neutral environment; pH = 7-14 - alkaline environment; pH = 1-7 - acidic environment [43; p. 235].

It is well known that the degree of acidity or alkalinity of drilling fluids has a significant influence on the manifestation of other properties. For example, by changing pH value, one can achieve certain rheological and filtration properties, inhibiting capacity of drilling fluids, their sedimentation stability, etc. [44; p. 311-313].

The pH value also affects the solubility of inorganic reagents (salts) and the efficiency (form of molecules) of polymer reagents. The optimum pH values are usually in the range of 9 to 11 [45; p. 66-68].

However, for alkaline solutions, the pH also increases with increasing pH:

disturbance of the stability of claystone well walls due to their additional wetting as a result of intensification of electro-osmotic processes [46; p. 85];

The chemical dispersion (peptization) of clayey rocks, which makes it difficult to remove them from the drilling fluid, thus causing an increase in its density, viscosity and static shear stress; [47; p. 140-141];

The natural permeability of productive sandstone reservoirs has been reduced

clayey reservoirs due to a reduction in the size of pore channels caused by the swelling of the clay component of productive formations, as well as due to clogging of these channels by clay particles migrating into them [48; p. 200].

Numerous scientific and technological studies and production tests indicate that the hydrogen index (pH) of drilling fluids should be in the range of 9-10 or more.

The thickness of the filtration crust indirectly characterises the ability of the solution to create a low-permeability filtration crust on the walls of the borehole [49; p. 181].

The formation of a crust on the surface of the borehole walls, depending on the type of drilling mud, must be for:

of lightweight drilling fluids - 0.5-0.6 mm;

of normal drilling fluids - 0.6-1.0 mm;

weighted muds - 1.0-2.0 mm.

Considering the developed requirements for creation of effective composite chemical reagents and drilling muds we have chosen objects for research. As a thickener and anti-filtration additive for drilling mud we have investigated: carboxymethyl cellulose (CMC), polyacrylamide (PAA), chemical reagents FHL-1 and SSB, Carboxymethylcellulose (CMC UP); to provide low value of water release and pH regulation caustic and soda ash; as structure-forming materials we chose nedopal, alumac, a waste product of Fergana-Azot chemical plant, and red clay, dolomite, hematite, barite, marble powder and others [50; pp. 174-176, 51; c. 36-34].

Table 2.1

Main physico-chemical characteristics of selected drilling powders and ingredients

Reagent	Specific weight, g/cm^3	Bulk weight, g/cm^3	Granulometric composition according to GOST 2851-45, residue on sieve No., %	Humidity, %	GOST or TU
CMC	(1,6 - 1,7) 1,66	(0,6-0,8) 0,674	№3-58, №8-5, №5-12,5, №10-5, №7-3,6, №2-15,9	10	TU5.588-2000
PAA	(1,302-1,31) 1,31	0,817	№3-82, №8-4, №5-2, №10-3,8, №7-2,4, №2-5,6	16-20	TU 6.1-00203849-29; 1994
FHL-1	1,052	0,596	№3-68, №8-3, №5-10,4, №10-4, №7-3,4,№2-11,2	12	TU 6.19.41-2008
Caustic soda	2,1	1,04	3-4 mm granules	1-2	GOST 2263-71
Calcine d soda	2,53	0,56	Powder up to 0,5 mm	1-2	GOST 5100-85
Alumac	2,86	1,18	№3-16, №8-3, №5-1, №10-14, №7-12, №2-54	1-2	

§2.2 Methods for obtaining samples of composite chemical reagents to determine their physico-chemical and technological properties

The methods we used to obtain and determine the physico-chemical as well as technological parameters of chemical reagent samples using nepal and other ingredients, as well as drilling muds based on them, are as follows. First of all, nepal and other ingredients undergo drying in the desiccator at 105±5^0 C for 2 hours until the residual moisture does not exceed 2%. Then they are dispersed on a shredder to particle size of 20-50 microns [52; p. 56, 53; p. 144-145].

The principle of operation of the shredder is as follows: the dried underflow is fed into the shredder, which consists of the inlet pipe 1, body 2, cover 3, inside of which there is a stationary disc 4 with the elements 5 concentrically placed around its circumference, movable (rotating) disc 6 and an electric motor 7 mounted on the shaft. Disc 6 on the end plane contains fingers 8 concentrically placed around the circumference of the disc. The disc fingers enter the gap between the circumferences of the concentrically placed segments. At the bottom of body 2 there is an outlet window 9. The shredder is mounted on a frame 10.

The shredding process is incomplete - from the spigot 1, the material entering the area of the rotating disc 6 is subjected to impact thanks to the element 5 and the rapidly rotating fingers 8. The under-shredded material exits through the window 9.

Samples of composite chemical reagents containing undopal are prepared as follows: 20 grams of undopal are weighed on an analytical scale with an accuracy of ±0.5 g. Then 80 g of powdered chemical reagent KPM-SK-2 on the basis of gossypol resin is weighed. The weighed components are thoroughly mixed in a mechanical mortar until a homogeneous powder is formed.

The rod 4 vibrates horizontally in an arc, which ensures good abrasion and mixing of the chemicals in cup 3. The rod 4 is equipped with removable weights 5, providing different degrees of abrasion of materials.

Principle of operation of the mechanical mortar: different chemicals are poured into cup 3 and mortar 4 is activated. After the composition has been thoroughly mixed and crushed, the mortar is stopped, the cup is emptied of the composition and the process repeats.

Drilling muds were produced in the rig shown in the schematic. The unit consists of a tank 1, an agitator 2, a frame 3, a valve 4, an agitator drive 5 and an inlet 6.

A sample is taken from the resulting composition for the preparation of drilling fluids. Depending on required density, viscosity, water yield, degree of shear stress and thickness of crusting the drilling mud is prepared on the rig. The process of preparation of drilling agents is as follows: the solvent is poured into charging tube 6, the powder is poured and with the help of rotating stirrer 2 the complete dissolution of the powder takes place. The prepared solution flows out of the tank 1 when valve 4 is opened.

The physical characteristics of the powdered ingredients and the physico-chemical and technological characteristics of the drilling fluids obtained from them are determined using standard methods, which are discussed below.

§2.3 Methodology for studying the physico-chemical and technological properties of composite chemical reagents

The properties of both inorganic and organic powder ingredients play a role in the production of drilling fluids based on them. Therefore, it will be necessary to study the properties of the powder ingredients first before obtaining quality drilling fluids. The bulk volume weight, specific gravity and particle size distribution of the powder ingredients play a special role in the production of quality drilling fluids.

In this regard, we present in this paragraph the methods we have chosen and used to determine the bulk volume weight, specific gravity and particle size distribution of the powder ingredients.

Methodology for determining the bulk density of powdery materials. Bulk volume mass $\gamma_{o.н}$ is the mass of a unit volume of loose powdery bulk material with voids.

To determine this index, a standard funnel is used, with a closure at the bottom. A preweighed vessel is placed under the funnel (g_1). The capacity of the vessel depends on the type and size of the material, e.g. for determining the bulk weight of sand - 1L (V_0). After opening the funnel closure, slowly fill the vessel to a height of 10 cm until a pyramid is formed, carefully cut off with a ruler and weigh the vessel with the material (g_2).

The bulk density is determined in g/cm^3 as an arithmetic average of the three determinations using the formula

$$Y_{o.н} = g / V_0 = (g_2 - g_1) / V_0$$

Methodology for the determination of specific gravity density of powdered ingredients. Solid phase density (specific gravity) is the weight - the mass of a unit volume of material in a dense state.

The density of solids in drilling fluids is determined using a volumetric meter (Le Chatelier-Candlau instrument), volumetric flask or by drying. The volumeter is a glass flask with an enlarged, funnel-shaped container in the middle part of the flask. The reservoir has a volume of 20 cm^3 and is marked at the top and bottom. The top neck of the flask is graduated in 0.2 cm^3 .

Fill the vial with paraffin up to the lower mark, place it in a vessel with water at 20°C and allow it to stand until the liquid in the vessel has reached the temperature of water (approximately 15-20 minutes). If the paraffin level changes with respect to the mark, either remove the excess paraffin with a strip of filter paper or refill it to the exact mark. Then in small portions a 100 g sample of the solid phase of the analyzed drilling fluid is poured into the volumeter, dried to a constant weight, grinded in a mortar and sifted through a sieve with 0.25 mm meshes. Weighing of the sample is made with an accuracy of 0.01 g as the arithmetic average of the results of three tests using the formula. Pouring of the sample is carried out with gentle shaking of the volumetric meter until the paraffin level rises to the upper mark or to a division above this mark within the graduated part of the pycnometer. The volumeter is then rotated around the vertical axis until no more air bubbles are released. The volumeter with a sample of solids is then held in a vessel with water until the paraffin level in the volumeter does not change.

The difference between the upper and lower marks of the paraffin level in the volumetric level gauge is used to determine the volume of paraffin displaced by the solid sample. The paraffin level in the volumetric volumetric level gauge before and after the pouring out of the sample is read on the bottom meniscus. The residue of the solids after filling of the volumeter is weighed again and the mass of the solids before analysis P_1 and the residue P_2 is used to determine the mass of the solids:

$$\mathbf{P = P_1 - P_2}$$

The density of the solid phase is calculated using the formula:

$$\mathbf{P_{TB} = P / V_K}$$

where P is the mass of solids poured into the volumetre, g;

VK- volume of paraffin displaced by the solid phase (solid phase volume), cm^3 .

While studying physical-chemical and structural-mechanical properties of drilling fluids we have mainly studied their density, static shear stress, viscosity, water yield and hydrogen index [54; p. 302-306, 55; p. 96-99].

To measure drilling fluid density we use areometer АБР-1, lever scales - density meter VRP-1, pycnometer, density meters AVP-1, PP-1, density indicator and others. For density testing of drilling fluids we used АБР-1 areometer. The areometer ADB-1 consists of the following main parts: a detachable weight, a measuring cup and a float with a rod (Figure 2.7). The measuring cup is attached to the float by means of pins. There are two scales on the rod: a basic scale to measure the density of the solution and a correction scale to determine the water correction. The instrument includes a bucket for water. The lid of the bucket serves as a sampler for the solution [56; p. 112-114, 57; p. 2-4, 58; p. 36-38, 59; p. 55, 60; p. 217, 61; p. 112].

The density of the drilling fluid in the case of mineralised water is calculated using the formula:

$$\rho = \rho \text{ main} + \Delta\rho ,$$

where: ρ - drilling mud density, kg/m^3 ;

ρbasic - density reading on basic scale, kg/m^3 ;

Δρ - correction (density readout on a correction scale), kg/m^3 .

The limit of admissible additional error when the temperature of the test solution changes for every 10^0 C, starting from $(20\pm 2)^0$ C - not more than 2 kg/m^3 , when the influence of climatic factors of the environment on the change in temperature of the test solution for every 10^0 C, starting from $(20\pm 2)^0$ C - not more than 10 kg/m^3 [62; p. 90-91].

Calibrate the areometer with pure fresh (distilled) water at $(20\pm 5)^0$ C. The reading should be 1 g/cm^3 [63; p. 41-43].

Structural and mechanical properties of drilling fluids are characterized by static shear stress. To measure static shear stress value, the

SNS-2 device is used, as well as rotary viscometers VSN-3, VSN-2M and FANN viscometer [64; p 425];

In order to assess the nature of the increase in structural strength over time, measurements are made after 1 minute (SNS_1) and 10 minutes (SNS_{10}) of rest [65; p. 32];

In addition to the aforementioned indicators, the structural-mechanical properties of drilling fluids are also characterised by the thixotropy coefficient [65; p. 32];

$$Kt = SNA_{10} / SNA_1 . \quad (2.7)$$

The required value of static shear stress after 1 minute (SST1, dPa) can be determined by the following formula [65; p. 32];

$$SNS_1 \geq 5\ [2 - echr\ (- 110\ d)]\ d\ (\rho_п - \rho\ ), \quad (2.8)$$

where: d is the notional diameter of the characteristic particles of the drilled rock, m;

$\rho_п$,ρ - rock and mud densities respectively, kg/m^3 .

The SNS-2 was used to measure the static shear stress [65; p. 32].

The SNS-2 consists of a case, a ladle, a screwdriver and six numbers of filaments, each with its own constant K in Pa/grad, as stated in the instrument's data sheet [66; p. 28].

The working principle of the device is based on measuring shear stresses in the controlled medium located between coaxial cylinders. The measure of shear stress is the angle of rotation of the suspended cylinder around its axis [67; p. 126].

Depending on the viscosity state of the test sample different filaments are used, the number of which determines the instrument coefficient. For filaments Nos. 1 and 4 the instrument factor is 0.043, for filaments Nos. 2 and 5 it is 0.12, for filaments Nos. 3 and 6 it is 0.3 [68; p. 71].

The static stress is calculated according to the formula:

$$\theta_{1,10} = A \cdot \varphi_{1,10} , \quad (2.9)$$

where: $\theta_{1,10}$ - static shear stress after 1 and 10 minutes, Pa;

A is the error factor (given in the appliance data sheet);

$\varphi_{1,10}$ - angle of twist of the filament measured after 1 and 10 minutes at rest, deg.

Viscometers VBR-1, VSN-3, VSN-2M, FANN and Marsh funnel are used **to measure viscosity of drilling fluids.** For determination of conditional viscosity we used viscometer VBR-1 [69; p. 124].

Conditional viscosity is a value indirectly characterizing hydraulic resistance to flow. In our country the conditional viscosity (CV, s) is determined by the time of flow of 500 cm^3 of flushing liquid through the vertical tube 2 of viscosimeter VBR-1 from the funnel 1, filled with 700 cm^3 of flushing liquid. The VBR-1 also includes a measuring cup 3 and a grid 4 [70; p. 18-19].

Check the viscosimeter's water figure by checking the flow time of 500 cm of pure fresh (or distilled) water^3 at $(20\pm 5)^0$ C. The elapse time shall be equal to (15 ± 0.5) s. If the value is greater than 15,5 s, the viscosimeter tube is cleaned, if less than 14,5 - the viscosimeter is replaced [72; p. 15-20].

When determining conditional viscosity under laboratory conditions, 200 cm of^3 solution is poured into a funnel and the time of flow of 100 cm^3 is recorded. The obtained value T=200/100 is multiplied by 4 [73; p. 51-52].

A VSN-3 rotary viscometer was used to determine plastic and effective viscosity of drilling fluids. At definition of plastic and effective viscosity the operation order is carried out as follows [74; p. 41]:

- agitates the mud at 600 rpm on the outer cylinder;

- use only two cylinder speeds: 600 and 300, or 400 and 200 rpm to obtain the values$\varphi_{1,2}$ and $n_{1,2}$.

The water yield of drilling fluids can be **determined** with the BM-6, UIV-2, FLR-1, ANI (ARI) and BAROID filter presses. Water yield of drilling agents has been studied with the BM-6. [75; c. 218-221].

The VM-6 consists of plunger 1, scale weight 2, cylinder 3 with screwed-in bushing 4, needle 5, filter cup 6, base 7, stopper 8, rubber gasket 9 and paper filter 10.

The appliance comes with a 0.5 litre oil reservoir and ash-free filter paper or prefabricated filters with a diameter of 70 mm.

The maximum water yield that can be measured directly on the BM-6 instruments is 40 cm^3 in 30 minutes. In order to be able to measure a higher value, double logarithmic grid forms are supplied with the instrument. The dependence of water yield on time on such a grid is expressed by a straight line [76; p. 36-38].

The thickness of the filter cake is measured. If a small error is neglected when the time is close to zero, then under conditions of static filtration of washing fluid through the filter paper the volume of filtrate is proportional to the square root of the filtration time: $\mathbf{F_{30} - F_0 = (F_t - F_0)[(30)^{0,5} /(t)^{0,5}]}$ (2.32)

where: F_t is the value of the filtration index after t minutes from the start of filtration, see[3] ;

F_0 is the magnitude of the error at near-zero time values, see[3] ;

F_{30} is the value of the filtration index for a standard measuring time of 30 minutes, see[3] [79; p. 366].

Errors at time values close to zero result from the ability of the smallest particles of washing liquid to pass through the filter paper before the pores are clogged. In measurements this is indicated by the BM-6 reading jumping from zero to a certain value, which is called the instantaneous filtration Fo (see[3]). After that only the filtrate passes through the filter paper [80; p. 366].

The use of dependence (2.32) for practical purposes allows significantly speeding up the process of determining the filtration index. Thus, if t = 7.5 min, then $F_{30} = 2 \cdot F_{7,5}$. Thus, for approximate estimation of filtration index for the standard measuring time F_{30} , it is sufficient to take a reading on the scale of VM-6 device after 7.5 min from the beginning of filtration and multiply it by two [81; p. 426].

In order to determine the filtration index, the values of which go beyond the scale of BM-6 device, as well as to predict the value of the filtration index by any arbitrary value of time since the beginning of filtration, the following formula, derived from formula (2.32) [82; p. 141], can be used.

$$\mathbf{F_{30} = F_t \cdot [5.477/(t)^{0,5}]} \quad (2.33)$$

It should be remembered that conditions of filtration crust formation at any accelerated method of filtration index determination by time do not correspond to standard ones, in connection with which its quantitative and qualitative characteristics are not indicative. At elevated temperatures, the water yield is determined using the unit [83; p. 56].

The thickness and permeability and permeability of the filtration crust are determined on the VIKA IV-2 device using a metal ruler with millimetre divisions or a caliper with a depth gauge. The thickness of filtration crust was determined on VIKA IV-2 [84; p. 10].

Let us consider the procedure of filter cake thickness measuring by the example of VIKA IV-2, which consists of a base 1, a stand 2 in which seat 3 moves freely, a scale with millimetre graduations 4, a screw 5, a holder 6, a pointer 7, a spring lever 8 and a pestle 9.

Determination of permeability of the filter cake. The permeability of the filter cake is the main parameter on which the rate of static and dynamic filtration depends. The permeability can be determined using the BM-6. For this purpose, after completing the measurement of the filtration index on BM-6, the glass of the device is washed with distilled water, carefully, without damaging the filtration crust. Having thus finished preparing the device for operation, distilled water is carefully poured into the filtering cup, having measured its temperature beforehand, the device is assembled and the same actions are performed as when determining the filtration index, or the time for which the weight scale will fall to a certain division is measured [85; p. 10].

The permeability of the filtration crust in nm^2 ($1nm^2 = 10^{-18}\ m^2$) is found from equation (34) reduced to the following form:

$$k = \frac{(V \cdot \delta \cdot \mu) \cdot 10^6}{\Delta p \cdot F \cdot t} \qquad (2.34)$$

where k is the permeability of the filter cake, nm^2 ;

V is the volume of distilled water in cm^3 that has passed through the filter cake in time t, s;

δ - thickness of the filter cake, mm;

μ - is the viscosity, distilled water at the temperature of the experiment, mPa-s;

Δp - pressure drop, MPa;

F - Filtration crust area, mm^2 .

Methodology for studying the hydrogen index - pH of aqueous-based drilling fluids determining their electrochemical properties. One of the main electrochemical indicators of aqueous-based drilling fluids is hydrogen ion exponent - pH [86; p. 35].

The hydrogen index describes the concentration of hydrogen ions [H+] in the drilling fluid (degree of acidity or alkalinity of water-based drilling fluids) [87; p. 17]:

pH = 7 - neutral environment;

7 < pH≤ 14 - alkaline environment;

1≤ pH < 7 - acidic environment.

The degree of acidity or alkalinity of drilling fluids has a significant impact on their other properties. Thus, by changing pH value one can change rheological and filtration properties, inhibiting capacity of drilling fluids, their sedimentation stability, etc. [88; p. 141-142].

The pH value also affects the solubility of inorganic reagents (salts) and the efficiency (form of molecules) of polymer reagents. The optimum pH values are usually in the range of 9 to 11 [89; p. 15].

However, for alkaline media, the probability increases with increasing pH:

The wellbore stability of clayey rocks is disturbed by additional wetting as a result of intensification of electroosmotic processes [90; p. 155];

The chemical dispersion (peptization) of clayey rocks, which makes it difficult to remove them from the drilling fluid, thus causing an increase in its density, viscosity and static shear stress [91; p. 78-91];

The natural permeability of productive sandy-clay reservoirs decreases due to a reduction in the size of pore channels caused by the swelling of the clay component of productive formations, and also due to the blockage of these channels by clay particles migrating into them [92; p. 48-49].

The colourimetric and electrometric methods are used to measure the pH value. The electrometric method, unlike the colorimetric method, is universal and more accurate (± 0.01 pH units). It is based on the ability of some substances to change the electric potential depending on the concentration of [H+]. For electrometric measurements special devices - pH-meters are used. The electrometric method, unlike the colorimetric method, is universal and more accurate (± 0.01 pH units). It is based on the ability of some substances to change their electrical potential depending on the concentration of [H+]. For electrometric measurements special devices - pH-meters are used. The operation of a pH-meter is based on the conversion of the electromotive force (e.m.f.) of the electronic system into a direct current, proportional to the measured value [93; p. 115-117].

The pH meter must be calibrated with buffer solutions before each use. Buffer solutions with a specific pH value are available for this purpose. At least two types of solution are used for calibration. For best results, the instrument must be calibrated with a buffer solution whose pH value is close to that of the test sample. The electrode should be left in the buffer solution for 5 minutes before measurement [94; p. 421-422].

The colorimetric method is based on the ability of some dyes to change colour depending on the concentration of hydrogen ions, and consists in determining the pH value using indicator (litmus) paper and reference coloured scales. This method is of low accuracy (± 0.5 pH units) and has a limited application (cannot be used to measure pH values of coloured liquids). We have carried out mainly the determination of water yield by colorimetric method [95; p. 11-12].

Thus, the following conclusions can be drawn on the basis of the analyses reviewed:

On the basis of comprehensive study of literary sources it is shown that drilling fluids used in oil and gas wells drilling have different requirements. In particular, their main technological properties include density, water yield, viscosity, static shear stress (SST), hydrogen index (pH), formation of filtration crust thickness [96; p. 87-89].

It has been established that each ingredient in chemical reagents and drilling fluids based on them is introduced for a specific purpose and performs its specific function. In particular, caustic soda, lime, alkaline ($Na_2 CO_3$) and acidic ($NaHCO_3$) salts, which have a certain concentration of hydrogen ions in the solution, are introduced to adjust the pH. Inorganic colloids (bentonite, asbestos) and, for oil-based solutions, organophilic clays and bitumens are used to impart thixotronous properties. The use of polysaccharides and synthetic water-soluble polymers is also very effective for these purposes. The most commonly used filtration reducers are inorganic colloids and organic - natural and synthetic high-molecular weight compounds of various chemical nature, such as carbon-alkali reagent, lignosulfonates, polysaccharides (starch, cellulose ethers) and acrylic polymers HIPAN, K-4, PAA, CMC and others [97; p. 110-111].

As diluents are used carbonate solution, lignin derivatives and inorganic reagents such as sodium hexometa-phosphate, sodium trisodium phosphate, sodium tripolyphosphate, etc. In addition, such reagents and materials are introduced into drilling fluids, which give special properties to drilling fluids, which include inhibiting and others [98; p. 265-267].

From the above data and their analysis one can see that during the last years the assortment of chemical reagents has increased sharply, new principles and methods of their treatment have been developed, but they remain expensive, scarce and ineffective, especially in highly mineralized formation water. In this regard, the task of creating effective import-substituting, non-deficient, accessible, relatively cheap and environmentally friendly chemical reagents for drilling mud stabilization on the basis of available raw materials remains very urgent [99; p. 50].

In view of the above, gossypol gum, a waste product of oil and fat production, CMC, PAA, PCL-1, caustic soda, soda ash, nitrogen fertilizer production waste - nepal, non-ferrous metal production and processing waste - alumac, red clay, dolomite, hematite, barite, marble powder have been chosen as research objects and their characteristics studied [100; p. 43].

Methods of obtaining and determining physical-chemical and technological properties of ingredients, chemical reagents, drilling fluids produced on their basis, using both fresh and mineralised formation water, have been established and described in detail [101; p. 196].

§2.4 Conclusions of chapter two

Thus, the choice and justification of objects and methods of research have been carried out, that during last years the assortment of chemical reagents has increased sharply, new principles and methods of their processing have been developed, however they remain expensive, scarce, ineffective, especially in highly mineralized formation water. In this connection the task of creating of effective import-substituting, non-deficit, accessible, comparatively cheap and environmentally friendly chemical reagents for drilling mud stabilization on the basis of available raw materials remains very urgent.

CHAPTER III. RESEARCH AND DEVELOPMENT OF EFFICIENT RAPA RESISTANT COMPOSITE CHEMICAL REAGENTS FROM ORGANOMINERAL INGREDIENTS BASED ON LOCAL AND SECONDARY RAW MATERIALS FROM DIFFERENT INDUSTRIES

§ 3.1 Study of the structure and physico-chemical properties of ingredients from local and secondary raw materials from various industries

Modern foreign and domestic experience opens up the possibility of extracting all valuable components from the oil-bearing raw materials processed in the republic, recycle wastes, produce a number of by-products both food and technical for various branches of the national economy. In this respect it is important to reveal and use all reserves of increase of technical level of the industry, to substantiate necessity of acceleration of introduction of the most important achievements of scientific and technical progress, especially the finished scientific developments, which allow to receive considerable economic effect.

In this section we present the results of studies of physical and chemical properties of such ingredients of organic origin. We investigated chemical reagent KCR-1 of domestic production, taken by us as a basis of the composition, carbonate-polymer slime (CPS), soda ash, caustic soda, various weighting materials.

KCR-1 is a ready-made chemical reagent of domestic production for preparation of drilling muds with reduced density. Technological characteristics of KCR-1 are given in table 3.1

The analysis of properties of KCD-1 has shown that when using this reagent for preparing low-density drilling muds the rheological properties of a solution keep stability during the increase of solid phase in the system.

Presence of surface-active, emulsifying and inhibiting properties due to composition, structure and interaction of ingredients of KCD-1 gives an opportunity to receive stable emulsifying system, which promotes to achieve high effect of drilling mud stabilization [Negmatova K.S. Dissertation. Chapter 4 page 125-130].

Table 3.1

Physico-chemical and technological characteristics of 10% drilling mud based on developed composite chemical reagents of KCR-1 type

Characteristics of drilling fluids	COHP-1-1	COHP-1-2	COHP-1-3	KCR-1-4	KCR-1-5
Appearance	dark brown powder				
Water solubility (10% aqueous solution)	soluble in water				
Density, γ, g/cm^3	0,83	0,84	0,85	0,86	0,87
Conditional viscosity of 10% aqueous solution according to SPV-5, T, s,	29	35	46	64	88
Water yield of 10% aqueous solution by the VM-6, V, cm^3 /30min	8,5	8	7	6	5,5
Static shear stress, SNS, 1/10min, mg/cm^2	18	23	28	34	41
Hydrogen index, pH	9	9	9	9	9
Crust thickness, mm	tracks	Traces	tracks	tracks	tracks

Study on the structure and chemical composition of sodium carboxymethyl cellulose (Na-CMC)

Carboxymethyl cellulose is a light to light beige crystalline powder, odourless and tasteless. Well soluble in water, alkalis; moderately soluble in acids, glycerine; insoluble in organic solvents, mineral oils. It is an ester of cellulose and glycolic acid.

When dissolved in water it forms viscous transparent solutions, which are characterised by pseudoplasticity and some by thixotropy. In aqueous solutions carboxymethyl cellulose is a weak surfactant (for 1% solution at

25°C 71 mN/m), it combines well with other water-soluble cellulose ethers, natural and synthetic polymers, as well as many salts of alkaline, alkaline earth metals and ammonium [102; p 74-76].

Blackening temperature 227°C, carbonisation temperature 252°C. Chemical formula: $[C_6 H_7 O2(OH)_{3-x} (OSH_2 SON)_x]_n$, where x = 0.08-1.5. Carboxymethyl cellulose is obtained as follows: α -cellulose (cellulose) obtained directly from plant fibres is soaked in strongly alkaline p and treated with chloroacetic acid, the resulting glycolate and sodium chloride are washed.

The most practical value is the sodium salt of carboxymethyl cellulose (Na-carboxymethylcellulose - Na-CMC), an amorphous colourless substance with a bulk weight of 400-800 kg/m^3 and a density of 1.59 g/cm^3. It is an anionic polyelectrolyte. It dissolves well in water. Viscosity of Na-CMC solutions is almost independent of pH. It binds water well, solutions are stable with respect to monovalent salts [103; p. 68-69].

Dry Na-carboxymethyl cellulose has little corrosive effect. It is biologically inactive and resistant to biodegradation, but its aqueous solutions undergo enzymatic hydrolysis during prolonged storage in air. The main property of carboxymethyl cellulose is the ability to form a very viscous colloidal solution, which does not lose its properties for a long time [104; p. 252-253].

Chemical properties: The molecular weight is 6400 (± 1,000) g/mol. Density is 0.5-0.7 g/cm^3 . It is insoluble in organic solvents.

With this in mind, we investigated the physico-chemical and technological properties of an aqueous solution of carboxymethyl cellulose of different concentrations of Fergana Chemical Plant furan compounds.

Table 3.2

Indicator	Value of the indicator
Degree of substitution	≥ 0,65
Viscosity, mPa	≥ 800-1200
PH	6,0 - 8,5
Humidity, %	≤ 10
Chloride content, %	≤ 7.0

Table 3.3 shows our control study on the physico-chemical properties of sodium carboxymethyl cellulose

Table 3.3

Physico-chemical characteristics of Na-carboxymethylcellulose

Name of indicators	Require ments of TSH 88.2-	The values obtained are		Method of control
		batch No.	batch no. 78	
Appearance	White to light cream coloured fine-fibre material			visually
Mass fraction of water, %, max.	12,0	6,7	7,3	according to GOST 16932
Degree of substitution by carboxylic groups, within	0,8-1,0	0,81	0,82	according to point 4.3.
Mass fraction of main substance in absolutely dry technical product, %, min.	50	52	52,7	according to point 4.3. TSh
Dynamic viscosity of an aqueous solution of carbocell with a mass fraction of 2% at 25°C, MPa, within	over 100	230	230	according to GOST 33
Solubility in water in terms of absolutely dry technical product, %, min.	97	97	97	according to point 4.3. TSh
Hydrogen index (pH) of an aqueous solution with a mass fraction of carbocell 1.5%, within	8-12	10,14	10,01	according to point 4.3. TSh
Degree of polymerisation, not less	700	700	827	according to point 4.3.

Na-CMC used in the study as shown in Table 3.3 had the following quality indicators: mass fraction of water - 8%; degree of substitution of carboxymethyl groups - 0.82; mass fraction of main substance in absolutely dry technical product - 58.5%; solubility in water in terms of absolutely dry technical product - 96.5%; hydrogen index of water solution - pH-11; degree of polymerization - 716.

Study of the structure and chemical composition of polyacrylamide

Polyacrylamide (PAA) is the common name for a group of polymers and copolymers based on acrylamide and its derivatives, the general formula is
(-CH2CHCONH2-)n [105; p. 25-28].

Synthesis. The main method for the synthesis of polymers based on acrylamide (AA) and other unsaturated amides is radical polymerization, which can be performed by all known methods: in mass of crystalline and melted monomers, in solution, emulsion and suspension. Each of the methods has its own characteristics which determine polymer properties and technical-economical indices of production. The regularities and technological aspects of the homopolymerization of AA and other unsaturated amides by different synthesis methods are discussed below. In addition, this chapter includes sections on the specifics of AA polymerization with other monomers and AA grafting to various polymers [106; p. 200-201].

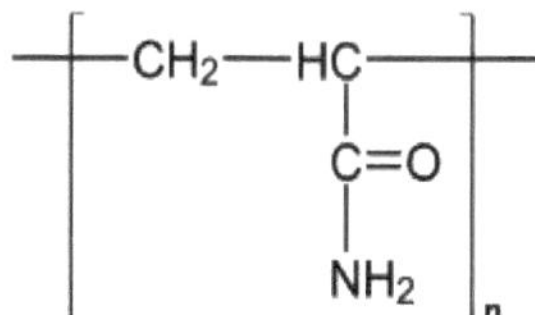

Elementary unit of a polyacrylamide macromolecule
Homogeneous polymerisation.

Homogeneous polymerisation refers to polymerisation processes in solvents in which both monomer and polymer are soluble. For polyacrylamide (PAA) the number of such solvents is small: water, formamide, acetic and formic acids, dimethyl sulfoxide (DMSO) as well as some water-organic mixtures. In addition, PAA has limited solubility in dimethylformamide (DMFA), ethylene glycol and glycerine. Polymethacrylamide (PMAA) is much less soluble than PAA. N,N-dimethyl- and N,N-diethylacrylamide polymers are soluble in water and insoluble in hydrocarbons. Poly-N,N-diethylacryl amide is soluble in acetone. Polymers with a longer alkyl substituent at the nitrogen atom are less soluble in water but more soluble in organic solvents. Poly-N-methyl- and poly-N-n-butyl methacrylamide dissolve well in DMF, N-(2-ethyl-hexyl)-methacrylamide polymer in toluene. Acryl- and methacrylurea polymers are soluble in concentrated solutions of hydrochloric acid [107; p. 14-15].

N-substituted acrylamides generally polymerise considerably faster than the corresponding methacrylamide derivatives. Acrylamides with bulk substituents, e.g. anthraquinone, do not undergo homopolymerisation [108; p. 29].

Among the methods of synthesis of PAA-based polymers an important place belongs to polymerisation in aqueous solutions. The main factors determining the prevalence of this method of polymerization are the high rate of polymer formation and the possibility to obtain under these conditions a polymer with a high molecular weight. It is supposed that a reason for the specific effect of water on PAA polymerisation is the protonation of the radical which leads to the localisation of the unpaired electron. As a result the reactivity of the macroradical which manifests itself in high values of the chain growth rate constant increases. Mutual repulsion of similarly charged radicals is responsible for limiting the rate constant of bimolecular chain breaking [109; p. 102].

In the non-potonated radical existing during polymerization in non-aqueous solvents, the conjugation of the unpaired electron with the -electrons of the C=O group leads to the stabilization of the radical and a decrease in its activity. Besides, high reactivity of PAA in aqueous solutions may be connected with the suppression of autoassociation of this monomer molecules due to the formation of hydrogen bonds with water molecules. In non-polar solvents, which are unable to form such bonds with PAA, the monomer is predominantly in an associated state in the form of cyclic dimers, trimers and linear multi-molecular associates. The data obtained in the polymerization of N-substituted acryl and methacrylamides are also consistent with this assumption. Thus, N-o-methoxy- and N-o-ethoxyphenyl-methacrylamides polymerize in bulk much faster than their m- and n-isomers, since the former have no molecular association, while the molecules of the latter are associated through hydrogen bonds [110].

In its turn, dimethylhydrazides acrylic acid (AA) and methacrylic acid (MAA), unlike their hydrochlorides, do not polymerize in mass because the molecules of these monomers are strongly associated. In aqueous solutions, however, both salts and free bases turn into polymers [111; p. 40-42].

Due to the above factors PAA has rather high value of ratio kp/k00.5 (according to different authors it is 3.2-4.4 for the temperature interval 30-60 °C) which along with low values of chain transfer constants for monomer and water determines the possibility of obtaining PAA in aqueous solutions with rate and molecular weight (MM) unattainable during polymerization in organic solvents. Other reasons for wide distribution of polymerization in water should be reduction of energy costs for the separation of initial monomer in crystalline form, which is also connected with the probability of its spontaneous polymerization, and for the regeneration of organic solvents, reduction of environmental pollution, and also exclusion of the dissolution stage of polymer reagents, used as a rule in form of aqueous solutions [112; p. 84-85].

Hydrolysed polyacrylamide (HPAA) is prepared by hydrolysis of paliacrylamide (PAA) in the presence of alkali or soda ash and sodium tripolyphosphate according to the method of W.L. Skalska.

Polyacrylamide is a liquid, usually of 8% concentration, well soluble in water.

For the preparation of hydrolysed polyacrylamide, 600kg of 8% PAA, 60kg of alkali, 60kg of sodium tripolyphosphate and up to 4m^3 of water are loaded into a clay mixer to give it a stabilising activity. Stirring is carried out until a homogeneous product is obtained. This reagent is called PC-2. It contains only about 1.5% dry matter. It is most expedient to use it to reduce water loss in non-mineralised drilling muds with low content of solid phase and non-clay natural suspensions [113; p. 524-525].

For this purpose it is more expedient to use PC-4, which is hydrolyzed PAA with addition of soda ash. Its optimum additives in drilling fluids range from 0.2-0.5% (in terms of dry matter) - for stabilisation of non-mineralised small-bodied washing fluids and up to 0.5-1.0% - for natural aqueous suspensions [114].

As PC-2 and PC-4 are of low concentration, they are not practical for use in reducing the watercut of heavy mineralised drilling fluids.

Hydrolysed polyacrylamide is very sensitive to polyvalent metal cations (Ca^{2+} , Mg^{2+} etc.).

When drilling wells with high bottomhole temperatures, treatment of non-saline drilling fluids hydrolyzed with polyacrylamide together with chrompic (0.01%) ensures maintaining low water yield and viscosity [115].

Hydrolysed polyacrylamide has no anchoring effect on clayey rocks. In order to obtain inhibited drilling muds after treatment with this reagent, sodium silicate should be added in the range of 2 - 5%. Satisfactory viscosity values of such systems are possible only at low content of clay fractions. Water yield of drilling fluids stabilized with hydrolyzed polyacrylamide at the addition of *2-5% sodium* silicate increases by 1-3 cm^3 for 30 minutes. Obtained systems are inhibited and can be successfully used for drilling unstable clay sediments, causing preservation of borehole size close to nominal size of drill bits [116; p. 70].

The polymeric reagents PAA and K-9 are used for the treatment of saline water. The precipitation of Na ions$^+$, K^+ and Cl^- occurs by reaction:

$$(K-9)n\begin{bmatrix} & H & \\ -CH_2- & C & \\ & CONH_2{}^+ & \end{bmatrix} Na + nCl^- \rightarrow$$

$$n\begin{bmatrix} & H & \\ -CH_2 - C - COO & & \\ & CONHCl & \end{bmatrix} Na$$

(PAA) n

$$\left[-CH_2 - \begin{matrix} CH - COOH \\ OC - C \equiv N \end{matrix}\right] + Na \rightarrow n\left[-CH_2 - \begin{matrix} CH - COONa \\ OC - C \equiv N \end{matrix}\right] + nH_2$$

Thus, carbonate wastes of $CaCO_3$, CaO, compounds of cationites and anionites as well as compounds of polymeric reagents such as PAA and K-9 are deposited in the bottom layer, which are conventionally called undopal [117; p. 131-132].

For the experiments to the solution of PAA having high sedimentation activity, capable due to the presence of electronegative oxygen atoms in the molecule to form sufficiently strong hydrogen bonds with hydroxyl groups of partially hydrolyzed PAA (Fig. 3.1). The formation of hydrogen bonds between the partially hydrolyzed PAA results in the formation of a binary composition with a dendrite-like structure, which effectively precipitates fine clay suspensions. When drying the binary composition, the patterns on the plates are preserved [118; p. 413-415].

The formation of hydrogen bonds between the two polymers can be judged from the change in IR spectra of the binary composition (Fig. 3.1.). In the region of 1000 - 1500 cm^{-1} a new "frequency band" of absorption bands appeared, indicating intermolecular interactions in the structured system [119; p. 252-253].

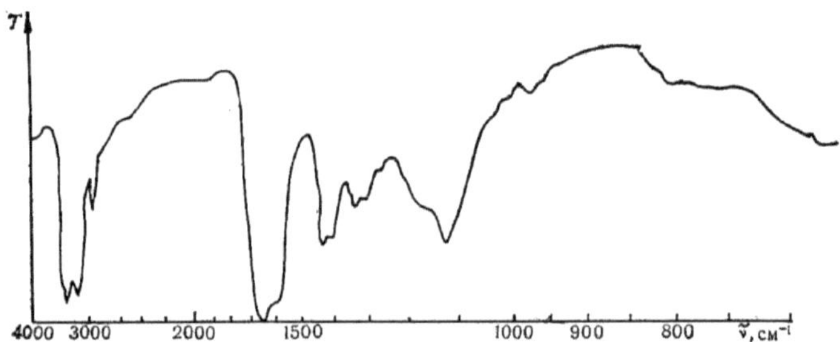

Figure 3.1. IR spectrum of polyacrylamide

Table 3.4

Physico-chemical characteristics of PAA - [-CN_2 CH ($CONH_2$)-]$_n$

Navoi Azot JSC

Name	Indicators
Acrylamide polymer content of the marketable reagent,%	
Grade A	≥50
Grade B	≥45
Ammonium sulphate content, %	
Grade A	≤38
Grade B	≤40
Insoluble sludge content,%	≤5
Product humidity, %	16-20
Colour	White, green, brown
Melting point,^{0} C	120
Dissolving time, 40^0 C	≤48
Chemical activity towards metals, oxygen, air and water	Close to zero
Electrical isolation during crushing, dissolving and transport	Not showing up
Fire risk, toxicity	Non-explosive, non-poisonous, fire-safe

Study of the structure and chemical composition of gossypol resin, a waste product from the oil and fat industry, for the production of composite chemical reagents.

The chemical formula for gossypol $C_{30} H_{30} O_8$ was derived by Clark (1927). Gossypol is a brown solid at room temperature, insoluble in water, soluble in most organic solvents: methanolEthanol, acetone, ethyl acetate, chloroform, phenol etc.. Gossypol exists in nature in two enantiomeric forms: (+) left-handed positive and (-) right-handed negative. In cotton plants gossypol is found as a mixture of both stereoisomers. Gossypol is a highly active chemical compound. The chemical structure of gossypol resin is shown on figure 3.2. [120; c. 60].

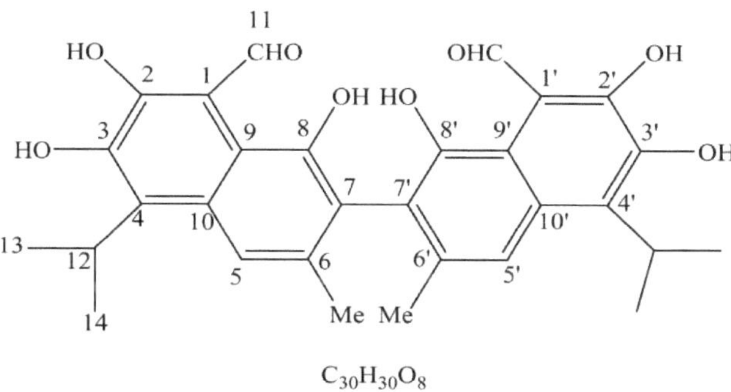

Figure 3.2. Chemical structure of gossypol resin

The unique features of the structure and chemical properties of gossypol molecules allow it to form fairly stable bonds with various proteins, as well as to easily integrate into phospholipid cytoplasmic membranes. Due to such interactions, gossypol blocks the activity of a number of enzymes participating in various metabolic processes in the cells of plants, microorganisms, and in the cells of many other species of living beings [121; p. 175].

Cottonseed oil, extracted from cotton seeds by the press method, has a dark colouring due to the presence of gossypol and its derivatives. It is therefore subjected to alkaline refining. The resulting precipitate - the primary residue - is treated with concentrated sulphuric acid and distilled.

During distillation of fatty acids extracted from cotton soapstocks, complex processes of further transformation of gossypol derivatives occur, in particular their interaction with each other with saturated fatty acids and other related substances. As a result, if fatty acids are distilled in a certain regime at 220-230^0 C, gossypol resin is formed simultaneously with fatty acids distillation [122].

According to the data of the republican oil and fats factories, during distillation of fatty acids extracted from cotton soapstocks the following parameters were obtained (from the amount of acids sent for distillation): fatty acids yield 75-85; cube residue yield 10-13; carbon monoxide at distillation 3.0-4.5.

Gossypol resin - contains usually 35-40% of products of oxidation, condensation, polymerisation and other reactions of gossypol, 40-50% of fatty acids and their derivatives and 10-12% of nitrogen compounds. It is a homogenous viscous mass of dark brown to black colour, insoluble in petroleum distillation products (petrol, paraffin, diesel fuel, chloroform, acetone, alcohols, esters, etc.).

Determining the composition of gossypol resin. Since 1958, when the use of black cotton soapstocks was first banned and the instruction was given to distill fatty acids for soaps during their processing, there has been an urgent need to study the composition of cotton soapstocks and cotton tars or gossypol tar, as it was called in the VNIIZh, and the search has begun for their rational use in the national economy.

Gossypol resin contains from 52 to 64% of hydrogenated fats and their derivatives, the rest are condensation and polymerization products of gossypol and its transformations formed during oil extraction, mainly in the process of distillation of fatty acids from soapstocks [123; p. 64-65].

Here is a typical composition of gossypol resin (Kattakurgan oil and fats plant) at 1st regime: 97.29% organic substances; 2.71% inorganic substances; 100% ester soluble substances; acid number 65.3 mg KOH; iodine number (Hanus) 99; saponification number 199 mg KOH/g; ether number 134 mg KOH; hydroxyl number 91%; 64% fatty acids liberated by saponification; 36% non-fatty substances; 0.2165% phosphorus (in terms of P_2O_5); 8.55% calcium in calcium salts of gossypol resin. The high ether numbers in gossypol resin indicate the presence of large amounts of lactones. Our studies have shown that gossypol resin contains 12% of nitrogen-containing compounds, 36% of gossypol transformation products, and 52% of fatty and oxyfatty acids, which was confirmed by IR spectroscopic studies (see Fig. 3.3).

As can be seen in the IR - absorption spectrum of gossypol resin - 1,1', 6,6', 7,7' - hexaoxy 3.3'-dimethyl - 5,5' - di-isopropyl-2,2'-dinaphthyl - 8,8'1 -

dialdehyde ($C_{30}H_{30}O_8$) detected frequencies at 3751, 3725, 3711, 3670, 3648, 3628, 3608, 3357, 2923, 2853, 1712, 1645, 1634, 1557, 1464, 1456, 1377, 1280 1110, 967, 842 and 723 cm^{-1}.

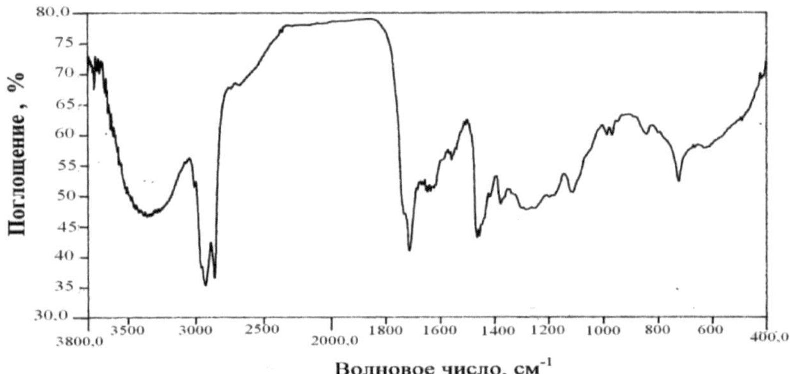

Figure 3.3. IR spectrum of gossypol resin

The main characteristics and properties of gossypol resin are shown in Table 3.5.

Table 3.5

Main characteristics and properties of gossypol resin

Properties	Indicators
Appearance	Viscous-fluid mass
Colour	Dark brown to black
Acid number, mg KOH	50-100
Ash content, %	1,0-1,2
Moisture and volatile matter content, в %	4,0
Solubility in acetone, in %	70-80
Specific weight, g/cm	0,98-0,99
Saponification number, mg KOH	80-130

Study of the characteristics of soda ash

Caustic soda is a chemical consisting of several elements: sodium, oxygen and hydrogen (one element each). Its formula is NaOH (figure 3.4).

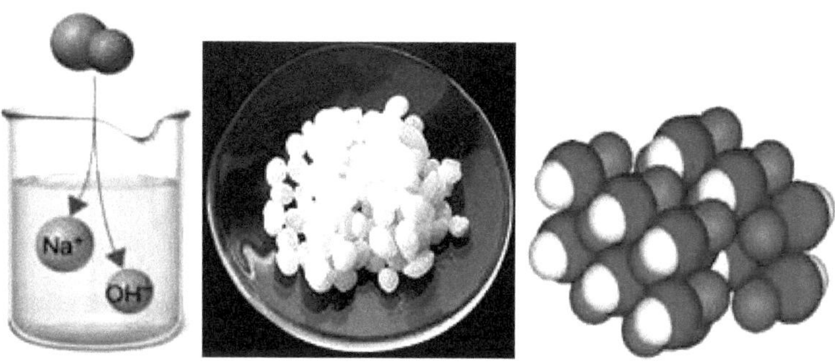

Figure 3.4 - Chemical formula and structure of caustic soda

The chemical formula for caustic soda is NaOH. It has the following properties:

- hygroscopic, in the air the granules spread out, absorbing water;
- dissolves in water, releasing large amounts of heat;
- does not react with plastic, rubber, steel, cast iron;
- contact with zinc, aluminium surfaces produces a violent reaction;
- Effectively dissolves grease and all organic matter: hair, paper, food residues;
- is volatile and should be kept in a tightly closed container.

Caustic soda is a strong poisonous lye. If its solution comes into contact with the skin, it can cause burns and sores. It is a class 2 substance, so care must be taken when using it:

- work wearing a mask, goggles, rubber gloves and overalls;
- ventilate the room well;
- Store in a closed container out of the reach of children and animals;
- If it comes into contact with the skin, neutralise with vinegar and wash the affected area with water;

- In case of contact with the eyes, flush with plenty of water.

In water NaOH dissociates almost completely into Na cations$^+$ and OH anions$^-$, increasing the pH of the solutions. Small additions of alkali cause dispersion of the clay, an increase in the electrokinetic potential and, consequently, a decrease in the viscosity and water yield of the clay solution. The high NaOH content in the clay mud can lead to an increase in viscosity and water yield caused by coagulation. The use of NaOH as a direct additive to the clay mud is therefore not advisable in some cases.

A study of the characteristics of soda ash.

Calcined soda (sodium carbonate or sodium carbonate). Na_2 CO_3 is a sodium medium salt of carbonic acid, a white powder with a specific gravity of 2.5 g/cm^3 [(GOST 5100-85)].

§3.2 Investigation of the physico-chemical and technological properties of the developed rapastable composite chemical reagents

Study of physico-chemical properties of water solution of carboxymethylcellulose of various concentrations**.** It should be noted, that at present the most widespread chemical reagent, used at preparation of drilling agents for drilling oil and gas wells and opening of productive horizons, is kaboxymethylcellulose (Na-CMC), received from cotton lint.

The effectiveness of drilling fluids is directly related to the quality of the chemicals used. Therefore, there are certain requirements for chemical reagents in terms of their physical-chemical and technological properties. In view of this, we studied physico-chemical and technological properties of aqueous solution of Na- CMC of various concentrations of domestic production. The Na-CMC used by us had the following indexes: mass fraction of water - 8 %; degree of substitution of carboxymethyl groups -

0.82; mass fraction of the main substance in absolutely dry technical product - 58.5 %; solubility in water in terms of absolutely dry technical product - 96.5 %; hydrogen index of water solution pH=11; degree of polymerization - 716 [124; p. 61-62].

Fig. 3.5 shows the results of the physical and chemical properties of an aqueous Na-CMC solution of different concentrations. From the course of the curves in Fig. 3.5 shows that the viscosity and density of solutions increase as the concentration of Na- CMC increases, while the water yield decreases. This pattern of the curves is explained by the fact that Na-CMC is a high molecular weight substance with a higher density than water. As it dissolves in water, the viscosity of the system increases, and with an increase in viscosity the water yield of the solutions decreases and vice versa [125; p. 50-52].

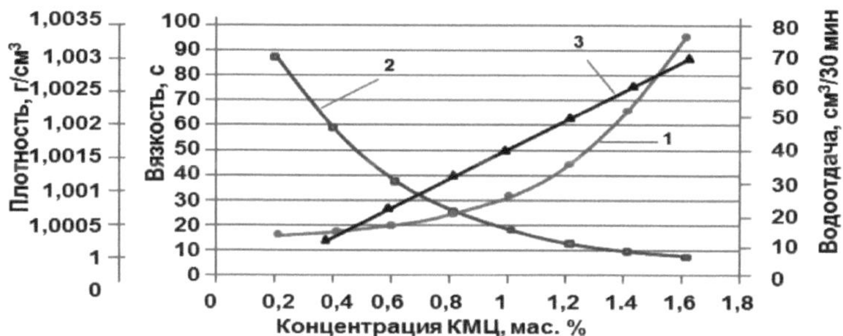

Figure 3.5. Viscosity (1), water yield (2) and density (3) of Na-CMC aqueous solutions versus concentration

The above is supported by the results of studies on the process parameters of aqueous Na-CMC solutions (see Table 3.6).

As can be seen from Table 3.6, the viscosity, density and water yield of the solutions significantly depend on the concentration of Na-KMC. The viscosity of Na-KMC solutions increases from 16 to 94 s. The density index changes insignificantly. However, the water yield decreases from 70 to 7 cm^3 /30 min. Thus, we can conclude that by selecting appropriate concentration of Na-KMC one can purposefully regulate rheological and technological parameters of the obtained drilling fluids, which creates favorable conditions for carrying out drilling operations [126; p. 88].

Table 3.6

Process parameters for aqueous Na-CMC solutions of different concentrations

Concentration of Na-CMC aqueous solution, wt%	Viscosity, c	Density, g/cm^3	Water yield, See^3 /30 min	pH
0,2	16	1,0004	70	7,5
0,4	18,22	1,0008	45	8
0,6	21,04	1,0012	30	8
0,8	25,08	1,0016	20	8
1	31,56	1,002	14	9
1,2	45,07	1,0024	11	9
1,4	65,84	1,0028	8	10
1,6	94	1,0032	7	10

Study of physical and chemical properties of drilling fluids using a modified gossypol resin powder stabilizer. In order to stabilize drilling muds produced on the basis of different chemicals we developed a chemical reagent on the basis of modified composite powdery gossypol resin, which was conventionally named KPGS [127; p. 323-325].

In addition to gossypol resin, KCHS contains caustic soda, soda ash and a hardener - alumac (which is a waste product of secondary non-ferrous metals), which help to transform gossypol resin into a water-soluble powdery state. Technical characteristics of GGBS are given above in Table 3.8 and in Fig. 3.6 [128; c. 235-236].

From the course of the curves in Figure 3.6 it can be noted that as the concentration of the CPGS in the aqueous solution increases, the viscosity (cr 1) increases from 15 to 23.8 s, while the water yield (cr 2) and density decrease. The water yield decreases from 26 to 8 cm^3 /30 min, and the density from 0.96 to 0.6 g/cm^3 . The latter is due to the fact that emulsion foaming occurs when dissolving CPGS in water, which leads to a decrease in density of the solutions [129; p. 48].

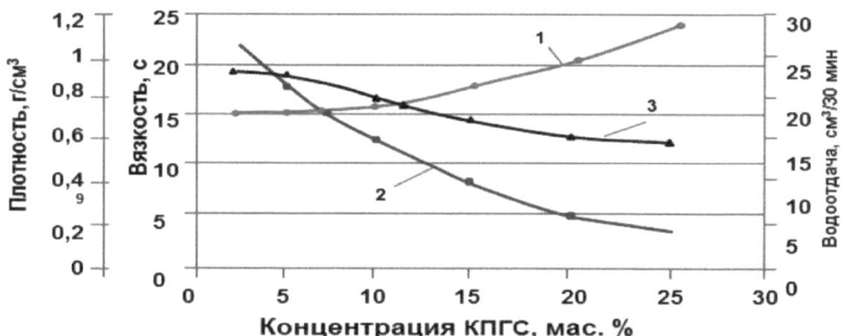

Figure 3.6. Viscosity (1), water yield (2) and density (3) of aqueous PPCS solution versus concentration

Table 3.7 shows the results of studies of the process parameters of aqueous PPCS solutions of different concentrations.

Table 3.7

Parameters of aqueous solutions of different concentrations of CHPSS

Concentration aqueous solution CBCP, wt%	Viscosity, C	Density, g/cm^3	Water yield, See^3 /30 min.	pH
2,5	15,05	0,96	26	7,5
5	15,1	0,94	21	8

7,5	15,3	0,92	17	8
10	15,6	0,88	14	9
12,5	16,3	0,82	12	9
15	18	0,76	10	9,5
20	20,3	0,68	9	10
25	23,8	0,6	8	10

From the data in Table 3.7 it can be seen that with an increase in concentration of CPGS from 2.5 to 25%, respectively, the viscosity of the solution increases, while the density and water yield decreases and the pH rises from 7.5 to 10 [130; p 4-7].

Thus, one can conclude from the combination of studies that the chemical reagent KPGS can be used to produce emulsions of different concentrations, used for drilling oil and gas wells with difficult technical and geological conditions [131; p. 7-8].

Study of technological properties of aqueous solutions of polyacrylamide of various concentrations. It is known, that polyacrylamide (PAA) is one of the components of composite chemical reagents, on the basis of which drilling muds for oil and gas wells are prepared. As our researches showed technological parameters of produced drilling agents with PAA directly depend on its concentration.

Figure 3.7 shows the viscosity, water release and density dependencies of aqueous PAA solutions on its concentration.

Figure 3.7. Viscosity (1), water release (2) and density (3) of aqueous PAA solutions vs. concentration

The curves in Figure 3.7 show that the viscosity and density of the solutions increase as the concentration of PAA increases, while the water yield decreases. This is explained by the fact that PAA is a high molecular weight substance with a higher density than water. As it dissolves in water, the viscosity of the solution increases and the water yield decreases, and vice versa. These conclusions are supported by the data given in Table 3.9 [132; p. 3-5].

Table 3.9

Process parameters for aqueous PAA solutions of different concentrations

Concentration of PAA aqueous solution, wt. %	Viscosity, c	Density, g/cm^3	Water yield, cm^3 /30 min	pH
0,2	17	1,0008	60	7
0,4	18,1	1,0017	40	7
0,6	19,4	1,0026	28	7
0,8	21,24	1,0035	18	7
1	23,6	1,0044	13	7
1,2	26,6	1,0052	10	7
1,4	30	1,0061	7	7
1,6	34	1,007	6	7

From the above data it can be noted first of all the following: the viscosity, density and water release values of the solutions depend significantly on the concentration of PAA. Thus, the viscosity of PAA solutions increases from 17 to 34 s. The density index, on the other hand, does not change much, and the water yield decreases from 60 to 6 cm^3 /30 min. On the basis of the results of studies it is possible to conclude, that by selecting an appropriate concentration of PAA, one can purposefully regulate technological parameters of drilling agents, obtained with its use.

Analysing the experimental data of dependencies of properties of aqueous solutions of Na-CMC and PAA on their concentration it is possible to conclude that the index of their density and viscosity tend to increase with the increase of their concentration. The water yield index in both cases tends to decrease. With the increase of Na-KMC and PAA concentration in aqueous solution from 0.2 to 1.8, the viscosity index increases from 15 and 17 to 95 and 34 s, respectively, and the density increases from 1.0001 to 1.003 g/cm^3 in case of Na-KMC and from 1.001 to 1.007 g/cm^3 when PAA is used. The water yield in the experiments decreases from 88 to 9 cm^3 /30 min, and in the case of PAA from 58 to 8 cm^3 /30 min.

Analysis of the dependence of the properties of aqueous PPCS solution on their concentration showed that the viscosity and density of the solutions tend to increase with increasing their concentration in the solution, while the water yield in both cases decreases. As can be seen from Fig. 3.6 and 3.8, with increasing concentrations of chemical reagents KPGS and FHL-1 in their aqueous solutions from 2.5 to 25%, density indices decrease from 0.9 to 0.6 g/cm^3 and from 0.99 to 0.88 g/cm^3 , respectively. The water yield is also reduced in the case of CPGS from 26 to 7 cm^3 /30 min, and in the case of FHL-1 from 25 to 4 cm^3 /30 min. The viscosity of the solutions increases in both cases from 15 to 24 s. From the above it follows that if drilling fluids with higher density, higher viscosity and lower water retention are required, Na-CMC and PAA should be used.

§3.3. Study of structure and chemical composition of mineral weighting agents based on local raw materials and waste products to produce composite chemical reagents used in the preparation of averaged and weighted drilling muds

Study of the structure and physical and chemical properties of clay minerals

In order to speed up the preparation of clay mortars, clays in powder form are mainly used.

Clay powder is dried and ground clay with or without chemical additives.

Clay powders are prepared from bentonite (PB), palygorskite (PP) and kaolinite-hydrosludite (HH) clays.

To improve the quality of clay powders and thereby increase the yield of the clay solution, a number of plants add various chemical reagents ($Na_2 CO_3$, M-14BV, methas, etc.) during clay grinding.

These clay powders are called modified clay powders (PPM, PPM). The yield of clay mortar from them is 1.5-2 times higher than from natural clay.

For example, the addition of $Na_2 CO_3$ promotes the conversion of Ca-bentonite into the sodium form, which swells better, hydrates more strongly and disperses more easily.

According to specifications GOST 39-202-86 clay powders depending on the mineralogical composition of clay raw material, the yield of clay solution with a certain viscosity and the presence of modifying additives are divided into the following types and grades (Table 3.10) [133; p. 116-118].

In the grades of clay powders, the first two groups denote the following: PB is bentonite powder, PP is palygorskite powder, PKG is kaolinite-hydrosludite powder. The letter M in the designation of grades of clay powders indicates that they are modified, i.e. contain chemical reagents introduced into them during milling. The last letters (A, B, C, G, D and H) in the designation of grades divide clay powders into groups according to their clay solution yield [134; p. 85].

Bentonite clay powder contains 70% or more of the mineral montmorillonite and is usually named after the clay deposit from which it is made. Its crystal lattice is three-layered, which has mobility, i.e. the ability to stretch and shrink due to binding water interlayers. Particles of bentonite clay have flake structure, but within elementary formations have crystalline structure. The average linear dimensions of the flakes are within 0.01-0.1 microns and are 10-100 times greater than their thickness. The specific surface area of bentonite is high and amounts to 450-900 m^3 /g. Swelling is reversible, though the swelling process can be delayed for 2 to 4 weeks. In this case the volume increase can be 20-fold [135; p. 237-238].

Bentonite clay powders containing mainly Na cations in the bulk complex$^+$, have a high exchange capacity (up to 1.5 mol/kg) and form suspensions with the required structural and rheological properties at low concentration of the solid phase [135; p. 239-240].

The quality of clay powders, especially those derived from clays with a calcium exchange complex, is improved by treating the clays with various reagents. Modified clay powder (MG) is bentonite treated with soda ash (up to 3-5%) and methas or M-14V (0.2-0.5%) at the grinding stage. It is possible to improve properties of mortar by treating it with reagents at the stage of preparation in the conditions of drilling. As a result of clay powder modification the mud yield can be 17-23 m^3 /t [136; p. 239-240].

Table 3.10

Types and grades of clay powders according to GOST 39-202-86

Mineralogical type of clay powder	Mark of clay powder	Mud yield Q, m^3 /t	Moisture of clay powder, %	The main rock-forming mineral
Bentonite	PBB	16	6- 10	montmorillonite
	PBV	12		
	UBG	8		
	BAP	5		
	NSP	<5		
	PBMA	20		
	BMPB	16		

	PBMV	12		
	PBMG	8		
Palygorskite	BCP	8	16-25	palygorskite
	PPD	5		
	PPN	<5		
	FPGA	12		
	PMG	8		
Kaolinite-hydrosludite	PCGD	4	3-8	Minerals of the kaolinite group, hydromica (illite) or both
	PCGN	<4		

Barite is a mineral species of variable composition (Ba, Sr) [SO_4], varying from the extreme barium member of barite proper - Ba[SO_4] to the strontium extreme member of celestine - Sr[SO_4]. The name barite comes from the Greek word barite - heavy [137].

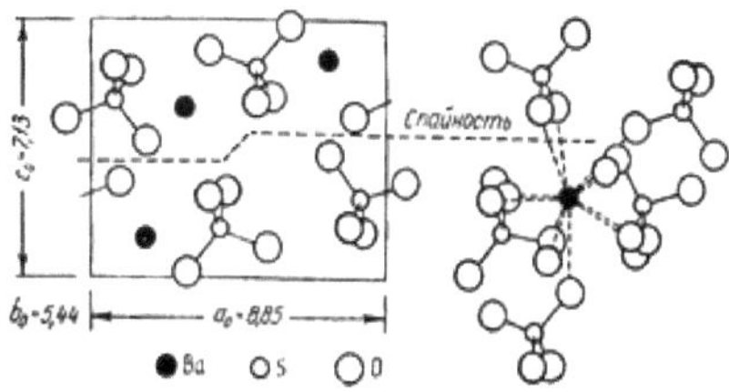

The chemical composition of barite is BaO - 65.7%, SO_3 - 34.3%, and celestine SrO - 56.4%, SO_4 - 43.6%. Ca, Pb and Ra are present as impurities.

Physical properties. The colour of barite is white or gray, sometimes red, yellow and brown (when stained with iron), as well as blue and greenish. The colour of celestine is bluish-white, bluish-gray, sometimes with reddish or yellowish tint [138].

Colourless, transparent crystals occur. The lustre is glassy, pearlescent on the cleavage planes.

Artificial production. Barite is obtained by precipitation by exchange decomposition of soluble barium salts and sulphates. Celestine is obtained by precipitation of solutions of Sr salts by sulphate solutions or H_2SO_4 , as well as by the action of excess HCl on precipitated $SrSO_4$ [139; p. 124-125].

Formation and deposits. Barite is a typical hydrothermal mineral in origin. It forms vein bodies in non-ferrous metal deposits. Barytes of exogenous origin are also known. Industrial deposits of baryte can be divided into the following main types: 1) hydrothermal (veined and metasomatic), 2) sedimentary, 3) alluvial [140; p. 284-286].

Dolomite is a mineral of the carbonate class of the chemical composition $CaCO_3$ -$MgCO_3$; dolomite is also called sedimentary carbonate rock, consisting of the mineral dolomite by 95% or more. It is named after the French engineer and geologist Deoda de Dolomieux (1750-1801), who described the features of dolomite rocks [141].

Chemical composition: CaO-30.4 %, MgO-21.7 %, CO_2 -47.9 %.

The CaO and MgO contents often fluctuate within small limits. Isomorphic impurities: Fe, sometimes Mn (up to a few percent), occasionally Zn, Ni, Co (in red dolomite, in Czechoslovakia (current Czech Republic) CoCO $content_3$ reached 7.5 %).

Bitumen and other foreign substances are known to be inclusions in dolomite crystals.

When the rock is ground to powder and dried in production conditions, dolomite flour is obtained. It is divided into 4 classes, within which 3 grades are distinguished.

The hardness of the mineral is medium, about 3.5-4. The density is 2.85 - 3.0 g/cm^3 . The mineral is strong, but rather brittle. It can be easily scratched with a steel needle [142; p. 94-95].

Dolomite is chemically similar to calcite. It differs from the latter by its more intense lustre and poor solubility. Definitely distinguish dolomite from limestone only by special chemical experiments [143; p. 124-126].

In the field, hydrochloric acid is used to distinguish between the two minerals. A small piece is placed on a glass surface and a drop of hydrochloric acid is added. If the substance boils and carbon dioxide is released, the mineral found is calcite. Dolomite does not react so actively with hydrochloric acid [144; p. 94-95].

There are several types of dolomite, the differences between which are due to their different natural occurrence. The minerals can be saddle-shaped, large, accreted, transparent, marbled, etc. They can be grey, white or pale yellow in colour, and less frequently stones can be black. The facets have a matte, pearlescent or glassy sheen [145].

Hematite - $Fe_2 O_3$ from the ancient Greek language. Amathitus - bloody by its thick red powder and trait colour. Its synonym is specularite. Hematite contains impurities: up to 11% TiO_2 , up to 14% $Al_2 O_3$, up to 8% $H_2 O$ (hydrohematite) and it comes in lamellar or tabular, less frequently rhombohedral crystals; solid earthy concealed crystalline masses, leafy and flakey aggregates, sintered formations; fine impurity in rocks and minerals, which stains them with intensive red colour (wax jaspers, red marbles, potassium feldspar, carnallite, sylvin, cancrinite, zeolites), etc.п. [146 c. 187-188].

The chemical formula of hematite shows that this ore is rich in iron, which is why the metal is mostly obtained from red iron. In terms of appearance, the rock has grey, dark, metallic and steel hues, often with inclusions of brown. Therefore the word "hematite" is translated from Ancient Greek as "blood-red". [147 c. 45-46].

The chemical formula for hematite given above reflects the content of the main substance. In fact, there are impurities such as titanium oxide and aluminium oxide. Ironstone also contains up to 8% water (chemically bound). The stable chemical formula of hematite is due to the fact that the main component of the ore ($Fe_2 O_3$) is a product of transformation of oxidized iron. The metal is bound to oxygen, so it is not subject to further corrosion"[148 p. 191-192].

There are several varieties of hematite, depending on their look and structure: Iron Mica is made up of scales with a distinct metallic sheen. Specularite also shines well in the light and is closer to silvery or grey shades of colour. Red glass head, also called bloodstone, is a stone with a brown, bloody hue. The iron rose really does resemble this flower because of its unusual structure. Red ironstone is only coloured in brown tones. The crystal is quite heavy due to its high density. Crystallochemical structure. Among oxygen ions in dense packing, in octahedral cavities between six oxygen anions there are cations of trivalent iron" [149; p. 193-194].

Two thirds of the octahedral cavities are occupied by cations. Each iron ion is surrounded by six oxygen ions, and each oxygen ion is associated with four iron ions [150; p. 210-211].

Marble powder (microcalcite) is chemically calcium carbonate $CaCO_3$. The appearance of the powder is white in colour. Marble powder is a raw material that has a crystalline structure, is resistant to acidic media and is less reactive than chalk (although chalk and marble are the same chemical composition), which opens up wide possibilities for its application as a way to improve product quality and reduce general plant costs.

Currently, marble powder attracts consumers with a combination of features such as: high whiteness; brightness; good optical properties; low oil absorption; resistance to external environment; practical absence of reactive impurities, at low cost.

Marble powder is widely used in various industries:

in the paint industry: water-based and solvent-based paints and enamels; industrial paints, printing inks, facade paints, powder paints, primers, road marking paints;

in the construction industry for the production of dry mixes, putties, mastics, sealants, adhesives, sealing compounds etc.

polymer industry: in the production of: plastic panels, PVC pipes, cables, cornices, PP and PE granules, shoe soles, synthetic leather, window profiles, etc.

in the ceramic industry

in the production of electrodes

in the carpet industry

in the production of animal and bird feed

in the production of detergents and household chemicals

in the paper industry (for coating and filling of paper stock)

rubber materials

in the plastics industry

it is recommended to use marble powder coated with organic acids as a filler, which gives the marble powder a hydrophobic property, creating a complete bond in an organic environment and easily dispersible in hydrophobic systems, has a high chemical resistance. These advantages are only achieved with a 100% hydrophobic coating.

in the oil and gas industry: in the construction of oil and gas wells, marble powder is used as a weighting agent for fresh and mineralised drilling muds up to a density of 1700 kg/m3 and also for the preparation of lighter cement slurries. Acid treatment increases its solubility to 98%. In drilling mud on the basis of marble, fractions from 0,7 to 10 micron are used: 0,7 micron, 0,9 micron, 2 micron, 5 micron. Our company also produces other weighting agent fractions to order.

Marble powder is sediment stable and does not precipitate even at low shear stress. It provides a high degree of recovery of the original permeability of the reservoir after acid treatment. It is used as a bridging (plugging) additive.

§3.4 Study of the composition and physico-chemical properties of plastic mineralised sinkholes and mechanisms of their interaction with composite chemical reagents in the drilling process

Salinity of water characterises its dissolved salt content in g/l, mg/l, g/m^3 , kg/m^3 . Formation waters always have a certain quantity (Q) of salts dissolved in them. Formation waters are divided into four groups according to their degree of salinity:

brines ($Q > 50$ g/l);

salty ($10 < Q < 50$ g/l);

brackish ($1 < Q < 10$ g/l);

freshwater ($Q \leq 1$ g/l).

Mineralisation of produced water increases with the depth of the reservoir. Mineralisation of oilfield waters ranges from a few hundred g/m^3 in fresh water to 300 kg/m^3 in concentrated brines.

Produced water contains ions of dissolved salts:

anions: OH^- ; Cl^- ; SO_4^{2-} ; CO_3^{2-} ; HCO_3^- ;

cations: H^+ ; K^+ ; Na^+ ; NH_4^+ ; Mg^{2+} ; Ca^{2+} ; Fe^{3+} ;

micronutrient ions: I^- ; Br^- ;

colloidal particles SiO_2 ; Fe_2O_3 ; Al_2O_3 ;

naphthenic acids and their salts.

Most of the water contains chloride salts, up to 80-90 % of the total salt content. In quantitative terms, the salt cations of formation waters are arranged in the following row: Na^+ ; Ca^{2+} ; Mg^{2+} ; K^+ ; Fe^{3+} .

The solubility of salts and the increase of their concentration in formation water is greatly influenced by temperature and the partial pressure of CO_2 .

The maximum solubility of $CaSO_3$ in water is observed at 0° C, it decreases with increasing temperature. The maximum solubility of gypsum ($CaSO_4$ - $2H_2O$) in water is observed at 40° C. With further increase in temperature it decreases. As the partial pressure of CO_2 increases, the solubility of $CaSO_3$ increases. Decrease of formation pressure intensifies the process of precipitation of CaCO salts₃ , etc. Changes in thermal and baric conditions in the formation, even at low salinity of formation water, affect the solubility of salts and the process of their deposition [151 p. 195-199].

Laboratory tests were carried out on initial clay solutions prepared on the basis of Shorsui red clay and highly saline formation water from the Berdach, Saule, Aral-4 and Balkan areas of the Ustyurt region [152 p. 15-17].

Chemical composition taken from above mentioned areas of formation water has been analyzed and the results of chemical analyses and influence of modified powder reagent - KPGS and rapaustic reagent KCR-RUS on technological properties of clay muds are given in table 3.11.

Table 3.11

Chemical composition of formation water in the Aral-4, Balkan, North Berdakh, Saule and Surgil areas of the Ustyurt region

Names of areas	The density of the densities	pH	Hardness mg/eq.	Cation content per litre mg			
				Naa^+	K^+	Sa^{+2}	Mg^{+2}
Aral-4	1,09-1,10	6,10	465,00	20695	600	700	5229
Balkan	1,11-1,12	6,60	223,00	39971	400	540	2383
North Berdach	1,06-1,08	6,5	131,00	14338	80	1320	990
Saule	1,07	5,30	124.00	12015	100	1560	559
Surgil	1,07	5,6	127,00	13604	186	1340	780

The table shows that aggressive cations Ca^{+2} , Mg^{+2} , K^+ , Na^+ are abundant in formation water.

§3.5 Comprehensive analysis of the obtained experimental studies and development of an optimal composition of rapastable composite chemical reagents from ingredients based on local and secondary raw materials from various industries

Rapa is indeed a very saturated salt solution, so saturated that it resembles a gel in consistency. It occurs not only in surface water bodies, but also deep in the earth's interior when rock salt layers are in direct contact with layers of water-saturated rock. Rapa (or rapaprojection) is considered a geological complication when drilling wells, as it greatly complicates the drilling process - the drilling tool slips and goes off the path [153; p. 82-83].

The brine consists of molten salts $MgSO_4$, MgCl and others, which are in molten form and are found in various depths of the green beds.

Experience of carrying out chemical treatments of drilling agents while drilling out chemogenic deposits has shown that the main complication is thickening and increase of water yield of drilling agents at the initial stage of opening of the upper salt pack. If the first one is explained by high concentration of clay and clayey rocks in the drilling mud, the second one is explained by presence of potassium and magnesium salts and table 3.12, which are strong coagulants, in the upper salt pack. While it is easy to get rid of the magnesium salts by introducing soda ash, it is almost impossible to remove the potassium salts from solution. For this purpose, as it has been shown before, the solution must be treated with lignosulphates to liquefy it (in particular with oxyl) and the water yield must be reduced with GRP like CMC, i.e. the treatment of the solution is similar to the treatment of potassium washing liquids. [3]Practice shows that thickening of solutions at salt extraction is not observed in those places where concentration of solid phase has been reduced to 30-35 weight % (γ=1,22-1,24 g/cm^00).

Solutions are pre-treated with oxyzyl in combination with CMC. Consumption of these reagents was 1% of the volume (Well 51 Urtabulak).

After cementing the intermediate casing, which overlaps the salt stratum, the formation is opened in most cases with saline mud, which is only in some cases diluted with fresh clay mud or process water.

Table 3.12

Results of chemical analysis of drilling mud leachates used in the drilling of the salt strata

Type of ions	Number of ions mg/l			
	No. 51 Urtabulak	No. 38 Dengizkul	№ 39 Chulkuwar	№ 53 Zevards
2	3	4	5	6
$Na^{+} + K^{+}$	40942	158612	51735	84387
Ca^{+2}	9420	2243.6	5012	3140
Mg^{+2}	7420	4331.7	3521	5760
Cl^{-}	96620	98630	95842	96821
SO_4^{-2}	6102	5049	6530	7350
HCO_3^{-}	488	442.5	501	472

Oxyl 0.5-1.5% as a 25% alkaline solution and high viscosity CMC grades 500-600 0.5-1.0% are added to this solution to determine filtration characteristics and viscosity. Sometimes CMC is combined with MK or Hypane. To reduce the density of the mineralized solution and to improve its quality 20-35% of oil is introduced into it, which is emulsified in the clay solution with the help of reagents oxycil and CMC. The parameters of the resulting solutions are in the range:

Density 1.15-1.17 g/cm^3

Viscosity 35-45c

Statistical shear stress through:

1 minute - 15-25 mg/cm^2

10 minutes - 35-50 mg/cm^2

Water yield according to BM-6 - 4.0-5.0 cm^3 /30min

pH - 9-9.5

Due to the fact that in some cases there are gypsum and anhydrite interlayers, as well as highly mineralised water, it is mandatory to add 0.1-0.3% NaOH and 0.3-0.5% $Na_2 CO_3$ into the clay mud. The purpose of the mentioned reagents is to remove Ca ions coagulating the mud . $^{+2}$

The emulsion saline flushing fluids used to penetrate productive horizons in the second group of fields are fully satisfactory for well intervention practices. The improvement of formulations in this case consists in that. That from one side it is suggested to use potassium washing liquids with low content of solid phase (clay) while drilling of paddy-salt deposits and in substitution of less effective emulsifier in saline solutions - CMC with more effective and cheaper one - modified starch in combination with caustic soda and soda ash. The formulations of oil emulsion solutions based on modified starch caustic soda and calcined soda, which are used for well intervention in saline sediments at the fields of the second group, have been developed by us as follows. Clay mud, saturated with formation salt and selected at a well with the following parameters: density - 1.28 g/cm^3 , viscosity - N/T, water yield - 6-7 cm^3 /30min, pH - 9.0, is diluted with technical water saturated with salt to density 1.20-1.22 g/cm^3 and viscosity 30-35% and processed with 3.0-3.5% of modified starch previously dissolved in water with added caustic. The NaOH to modified starch ratio is 1:10.

Once the starch reagent has dissolved, it is stirred into the clay solution as a 10% solution. Once the starch reagent has been dissolved, the oil is added while stirring vigorously. The amount of oil introduced varies from 20 to 50% of the volume, depending on the required density of the solution. Parameters of the oil emulsion solution obtained are as follows: density -

1,15-1,17 g/cm^3 , viscosity - 50-55s, SNS 15-20/30-40 mg/cm^3 water yield - 2,5-3,0 cm^3 /30min, pH 9,0-9,5.

Heating the above solution in a consistometer at 900C and 200kg/cm pressure^2 for a couple of hours with periodic stirring resulted in the following changes of parameters: density 1,16-1,18g/cm^3 , viscosity 40-45s, SNC 10-15/25-30mg/cm^3 water yield 3,0-4,0 cm^3 /30min, pH 8,5.

Oil emulsion mud tested under downhole conditions with UIB-2 device has shown that its water-releasing capacity under 25 kg/cm pressure fluctuation^2 is 25 cm^3 and thickness of indelible crust is 3,5 mm. The water yield under similar conditions was 42cm^3 in 30 minutes and the indelible crust thickness was 5mm.

The obtained data prove that starch reagent is more effective emulsifier for saline clay oil emulsion solutions than high viscosity CMC. Thus, it follows that joint drilling out of salt-and-anhydrite strata and productive formation in the fields of the second group can be carried out with oil emulsion clay solutions using modified starch and caustic soda as emulsifier.

The process of joint drilling of salt strata and pay horizon in the fields of the third group with abnormally high formation pressure provides for further deepening of the boreholes in the up-gradient salt-saturated clay mud. Proceeding from the above-mentioned we offer new developed composite chemical agents KCD-1RUS, KCD-2RUS, KCD-3RUS and KCD-4RUS for preparing rapastable stable drilling mud with density 2,1-2,2 g/cm^3 . At that barite and hematite from Bekabad metallurgical combine wastes are mainly recommended as weighting agent.

On the basis of results of laboratory and bench tests on producing rapastable drilling muds on the basis of rapastable composite chemical reagents KCD-1RUS and KCD-2RUS a number of compositions of composite chemical reagents have been developed, which are shown in tables 3.13 and 3.14.

Table 3.13

Technological parameters of rapastable weighted drilling mud based on KCR-1RUS, KCR-2RUS and barite

Composition of a rapable weighted drilling mud					Technological parameters				
KHR-1RUc , %	KHR-2RUc , %	Barite masses, h	NaCl, %	Oil, %	γ, g/cm 3	T_{500}, s	B, cm^3 /30 min	K, mm	pH
10	10	100	-	8	1,63	62	0-1	0,2	10-11
10	10	150	-	8	1,84	105	1-2	0,4	10-11
10	10	200	-	8	2,03	265	2-3	0,6	10-11
10	10	200	15	8	2,09	136	4-5	0,6	8-9
Heating at 100^0 C for 2 hours					2,09	88	5-6	0,6	8-9

One can see from the table, that the weighted drilling muds based on rapastable composite chemical reagents KCD-1RUS, KCD-2RUS with barite, prepared with fresh and salt water, allow getting drilling mud with density from 1,63 to 2,09 g/cm^3 and water productivity in the range 1-6 cm^3 /30 min.

Further we consider technological characteristics of drilling fluids based on rapastable composite chemical reagents KCD-1RUS, KCD-2RUS with hematite. Tables 3.14 show the results of research of technological characteristics of drilling fluids based on KCR-1RUS, KCR-2RUS and hematite.

Table 3.14

Technological parameters of rapastable weighted drilling mud based on KCR-1RUS, KCR-2RUS and hematite

Composition of a rapable weighted drilling mud	Technological parameters

KHR-1RUc, %	KCR-2RUc, %	Hematite, wt.%	NaCl ,%	Oil,%	γ, g/cm³	T500, s	B, cm³ /30 min	K, mm	pH
10	10	100	-	8	1,64	66	0-1	0,25	10
10	10	150	-	8	1,78	76	0-1	0,3	10
10	10	200	-	8	2,03	82	1-2	0,4	10
10	10	250	-	8	2,10	262	2-3	0,7	10
10	10	250	15	8	2,17	242	3-4	1,6	9
10	10	250	30	8	2,21	126	4-5	0,8	8-9
Heating at 100⁰ C for 2 hours					2,21	87	5-6	0,8	8-9

Proceeding from above-mentioned results, it is possible to draw a conclusion, that new rapastable composite chemical reagents are suitable for production of weighty rapastable drilling muds with density from 1.64 to 2.21 g/cm^3 , pH 9-1, viscosity 66-262 s, water yield 1-6 cm^3 /30 min at temperature 20-100^0 C and mineralisation with NaCl up to 30%, that it is reasonable to recommend for production testing.

Therefore, starting from above-mentioned laboratory tests and technological indicators, it is possible to conclude that based on the positive results obtained, new composition chemical reagents KCD-1-RUS (KCD-1+barite), KCD-2-RUS (KCD-1+hematite) are good for obtaining heavy-duty rapastable drilling muds, KCD-3RUS (KCD-2+barite), KCD-4RUS (KCD-2+hematite) are suitable for producing weighted rapastable drilling agents, that is why it is recommended for production testing and their use in the process of oil and gas wells drilling in case of rapastable drilling [136-140].

§3.6 Conclusions to chapter three

1. The paper studies physical and chemical properties of rapastable composite chemical reagents and develops their optimal compositions for producing weighted rapastable drilling muds.
2. Stabilising properties of rapastable composite chemical reagents and their optimum contents in the composition of weighted rapastable drilling

muds are determined.

3. The regularities of physical and chemical properties of rapastable composite chemical reagents and salt-resistant drilling muds in weighting agents have been revealed.

CHAPTER IV. RESEARCH OF PHYSICO-CHEMICAL PROPERTIES AND DEVELOPMENT OF TECHNOLOGY OF COMPOSITE CHEMICAL REAGENTS AND RAPASTABLE DRILLING MUD PRODUCTION

§4.1 Study of the physico-chemical properties and technological characteristics of composite chemical reagents

To study the mechanism of action of composite chemical reagents CMC, KCR-1 and series KCR on malodorous drilling mud we studied adsorption on quartz and clay powder, change of surface tension of solution before and; after adsorption.

The results obtained show that the adsorption of CHP-RUS type on quartz is characterized by growth up to reaching 10% of its concentration after which the value of adsorption decreases and remains constant at 17.5% -concentration. This isotherm of adsorption corresponds to the Langmuir monomolecular isotherm, when the solute is adsorbed not over the entire surface of the adsorbent, but only on its active centres. At low concentrations (0.1% and 0.25%) molecular ions are adsorbed on quartz, starting from 0.5% concentration (region of BÇA formation) - micelles. Adsorption of anion-active CCR-RUS on negatively charged quartz seems to be explained by the difference between the charge densities of CCR-RUS macromolecules (carriers of high negative charges) and quartz particles (carriers of low negative charges). In this case there is a possibility of fixation of CXR-RUS due to the formation of hydrogen bonds between groups of molecules with polar groups on the surface of quartz particles [154 p. 207-208].

We carried out studies to determine the surface activity of anionic CCR-RUS in comparison to sulphanol and OP-10. These studies showed that CPM already at a concentration of 0.1% reduces the surface tension of water by a factor of 2. By the ability to reduce surface tension CCHR-RUS surpasses OP-10, but by the results obtained it is close to sulfanol (Table 4.1) [155 p. 534-538].

To study the effect of synergism we studied the variation of surface tension of aqueous solutions of KCR-1RUS and sulphanol without and with the addition of K-4. The study showed that the addition of K-4 contributed to a sharper reduction in the surface tension of sulphanol than for our synthesized preparation.

Table 4.1

Surface properties of KCR-1RUS and known surfactants

Surfactant concentration	Surface tension, 5 10~3, n/m		
	KHR-1RUc	Sulphonol cation	OP-10
0,01	60,7	61,2	50,9
0,05	42,1	42,6	41,6
0,10	38,0	38,6	41,0
0,25	36,2	36,9	36,1
0,5	35,9	36,2	35,6
1,0	34,8	35,1	35,5

The study of adsorption by changing the surface tension of the solution of CCP-1RUS on clay and loam showed that with increasing concentration of polymer-containing reagent, the surface tension decreases uniformly. CXR-1RUc adsorbed more on clay than on loam at equal concentrations of polymer-containing reagent. This appears to be due to the mineralogical composition of the adsorbent [156 p. 255-257].

Based on the results of the studies it can be concluded that the surface-active agents KHR-1RUS, designed to produce direct emulsions of oil type in water, are polymeric reagents of semicolloid type, capable of forming true molecular solutions at low concentration and colloidal, thermodynamic stable micellar solutions - at high concentrations. Filtration kinetics of fresh and saline clay solution based on Shorsui clay solution with addition of KCR-1 reagent are shown in Figure 4.1 and 4.2.

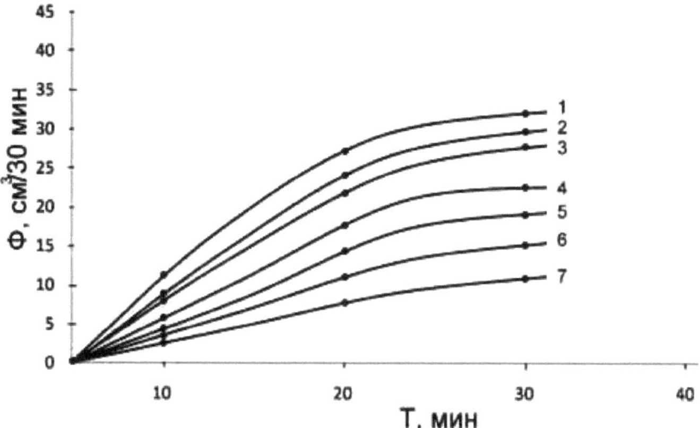

1 - no additive; 2 - 0.5 % KMC-500; 3 - 1.0 % KMC-500; 4 - 0.5 % K-4; 5-1.0% K-4; 6-0.25% KHR-1RUS; 7-0.5%

Figure 4.1 - Filtration kinetics of Shorsui clay mortar with various additives

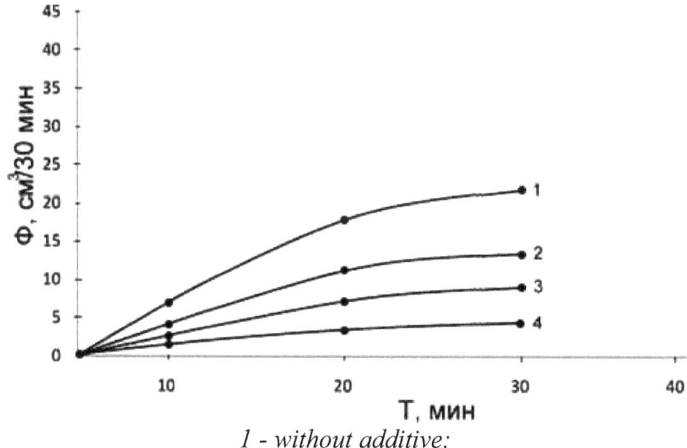

1 - without additive;
2 - 0,5 %: 3 - 1.0 %; 4 - 1.5 % KCR-1RUc.

Figure 4.2 - Filtration kinetics of saline Shursui clay solution with the addition of KCR-1RUS reagent

The analysis of research works, directed to creation chemical reagents for drilling agents and improvement of their operational properties, shows that at the present time a great number of chemical reagents, capable to be used in different geological conditions, has been created. However, problems connected with the manifestation of "RAPA" in different fields have not been practically solved. Particularly in our republic this problem is acute because of different depths of saline strata, which are the main reason of "RAPA" manifestation.

Our developed composite chemical reagents (CCR-RUS) using waste products of various industries and local raw materials contribute to the solution of this problem and intensification of oil and gas production, hence increasing their production volume.

We have developed a composite chemical reagent KCR-2 on the basis of KCR-1 using carbonate-polymer sludge from PO "Ferganaazot". [157 c. 43-47].

The composition of carbonate-polymer slurry and gossypol resin is a multi-component system and contains structure builders, peptides and polymer reagent-polyacrylamide. Carbonate-polymer slurry is a homogeneous highly dispersed mass of white-matte colour, well hydrated in water and quickly soluble in hydrochloric acid, has a hydrogen index pH≥12, density ρ = 3,05 - 3,18 g/cm^3 , and gossypol resin is a surface active substance, which gives it rapproof.

Laboratory tests on drilling mud treatment have been carried out. The results showed that CXR series improves parameters of drilling mud and practically does not concede to parameters of mud treated with widely used as: USHR, CMC, K-4 and Hypane. A new reagent has been investigated in laboratory conditions. It is a gossypol resin, which has got technological properties influencing on the improvement of drilling mud parameters along with already known and widely used reagents and is easily available and practically cheap. According to all the data of laboratory researches it can be recommended for implementation in drilling wells in the field "Plateau Ustyurt", with rapa-appearances.

Studies to determine the effect of inorganic salts on the stability of inverted emulsion based on the CXR series showed that the maximum saturation of the aqueous phase with salt is 20% of the volume of the emulsion taken. Therefore, in a further study of the effect of drilled salts on emulsion properties, the maximum additions did not exceed this value. The analysis of obtained data on changing values of technological parameters of SER (density, conditional viscosity, static shear stress, filtration, electrical stability) shows that getting sodium chloride in the emulsion does not essentially worsen its quality indicators and especially its stability. This is due to a significant reduction of ion transfer numbers and their coagulation ability in the emulsion medium compared to conventional water-clay solutions.

We investigated [137] the possibility of weighting the emulsion solutions based on CHR (different composition) by introducing barium sulphate with a density of 4.1 g/cm^3 . The experimental data obtained show that the order of barite introduction into emulsion does not essentially influence its stability, and the emulsion solutions themselves are weighted with flotation barite up to specific weight of 2,5% g/cm^3 . It is noteworthy that the addition of barite in very large quantities does not impair the stability of the emulsion. This is probably due to the modification of the surface of the weighting agent, adsorbed by surfactant molecules to form very strong organomineral films on the interface between the two phases. It is not excluded that the smallest particles of barite act as a solid emulsifier, increasing the stability of GER. Significant difference between obtained weighted solutions and solutions prepared on the water basis is their reduced filtration and weak colmatation ability. The thickness of the resulting crust is an order of magnitude smaller than that of the water-based solution.

All of these data together suggest that emulsion solutions based on the CXR series are highly effective as drilling fluids.

The study of rheological properties of HER, based on a series of KCR, on a rotary viscometer (plastometer - 2) at normal division in the temperature range 283-363K showed that the rheological behavior of the emulsion is not qualitatively different from the rheological behavior of water-based clay solutions. The main difference is that oil-based solutions have, under normal conditions, higher plastic viscosity and lower values of static shear stress (yield stress) than water-based solutions. Values of rheological parameters of emulsion solutions of different compositions at different temperatures (table 4.2) show, that plastic viscosity of emulsion solutions, both weighted and without weighting agent additives, tends to decrease with increasing temperature. Moreover, the viscosity decreases by several orders of magnitude when the temperature rises in the range 283 - 263 K.

The above behaviour of emulsion solutions is related to the features of the interfacial surfaces and is common to highly concentrated emulsions.

The study of rheological properties of HER on the basis of KCR-1 allows one to conclude that highly stable drilling muds can be obtained on their basis.

Table 4.2

Rheological parameters of a reversed emulsion

Reverse emulsion formulation	Rheological parameters	
	303 K	353 K

	η/ rush	τ,10^{-3}	η / rush	τ, 10^{-3}
Petroleum 40%, water 60%, KCR-1-1 (4:1), clays 5%	38	60	12	20
Composition 1 + 60% hematite	98	350	28	110
Oil 40%, water 60%, clays 5%	50	20	14	17

Studies of the filtration properties of aqueous reversed emulsions have shown that the filtration capacity of HERs is usually a whole order of magnitude lower than that of water-based solutions (Table 4.2). This phenomenon is explained by the clogging effect of oil droplets (straight emulsions) or water (HER) which are pushed into the capillary channels of the filter cake during filtration. Filtration of an emulsion based on our developed HER-1 did not exceed 3 cm^3 in 30 minutes and tended to increase at higher pressure and temperature drop. Reduction of viscosity of HER based on KCD-1 was achieved by us adding oil (diesel fuel) and reducing the solid phase in water. The static shear stress is increased by introducing organophilic clay powder (chalk) into the emulsion or by increasing the water content. The reduction of this parameter is done by the same method as the viscosity parameter. A study of the filtration of reversed emulsions found that a reduction in the water yield index is achieved by the introduction of organophilic clay, chalk and barite into the emulsion, as well as the KPCS itself.

Geological cross-section of Middle to Upper Jurassic deposits. Purpose: production-gas well design:

shaft direction 530mm - 9m

426 mm long directional extensions - 50 m

conductor 298.4 mm - 400 mm

intermediate column 219.1mm - 1,500m

139.7mm production casing - 2950m

Table 4.3

Influence of modified powdery reagent - KPCS and composite chemical reagent - CCHR on the technological properties of clayey drilling muds

№	Composition of clay mortars	Technological parameters				
		ρ, g/sm^3	T_{500} , s	B, sm^3 /30min	K, mm	pH
1	1 litre layer, Aral water - 4 + 400 g Shorsu kr. Clay powder	1,22	20	40>7.5 min.	10	6
1.1	No. 1 source, rp +10% CPCS	1,20	32	16	1,5	11-12
1.2	No. 1 outcome, rp +10% CXR-1	1,17	140	2	0,5	11-12
1.3	No. 1isoc, rp +10% CXR-2	1,15	200	2	0,5	10-11
1.4	1 litre plastic, water dense. Balkan +400g Shorsu kr. Clay powder	1,30	17	40>7.5 min.	10	6
2.1	No. 1 source, rp +10% CPCS	1,19	24	20	2,0	10-11
2.2	No. 1 outcome, rp +10% CXR-1	1,20	100	7	0,8	9-10
2.3	No. 1 outcome, rp +10% CXR-2	1,20	180	1,5	0,5	11
2.4	1 litre layer, water density North Berdach+ 400 g Shorsu kr. Clay powder	1,23	20	40>7.5 min.	10	6,5
3.1	No. 1 source, rp +10% CPCS	1,03	36	8	0,5	10-11
3.2	No. 1 outcome, rp +10% CXR-1	0,9	60	4	0,3	10-11
3.3	No. 1 outcome, rp +10% CXR-2	0,93	N.t.	4	0,3	10
4.	1 litre of plastic, kinda dense. Saule + 400 g Shorsu. clay powder	1,23	20	40>7.5 min.	К)	5,5
4.1	No. 1 source, rp +10% CPCS	1Д7	20	28	3	10

4.2	No. 1 outcome, rp +10% CXR-1	1,16	52	9	0,8	10
4.3	No. 1 outcome, rp +10% CXR-2	1,15	N.t.	5	0,3	10
5	To solution no. 4.2 +80% Barite	1,60	172	12	1,5	10
6	To solution no. 4.3 +80% Barite	1,60	140	10	1	10
7	No. 4 +40% Shorsu kr. Clay+10% CXR+20% NaCl	1,32	56	5	0,5	12

§4.2 Investigation of the physical-chemical and technological properties of drilling fluids using newly developed composite chemical reagents

To develop rapastable composite chemical reagents KCR-RUS we have studied basic parameters of weighted drilling muds using selected reagents and organomineral weighting agents as density, viscosity, filtration index, static shear stress, stability, sedimentation index, hydrogen index. For this purpose we have preliminary investigated technological characteristics of CCH, which was accepted by us as a stabilizing basis for development of rapastable composite chemical reagents for weighted drilling muds.

Further, the physico-chemical properties of CCHR-based drilling fluids were investigated.

In studying physico-chemical properties of the developed composite chemical reagent its aqueous solutions were prepared using fresh water. Preliminary researches on determination of dependence of water release and viscosity on concentration of CCHR without weighting agents were carried out and the following results were obtained which are shown in figure 4.3.

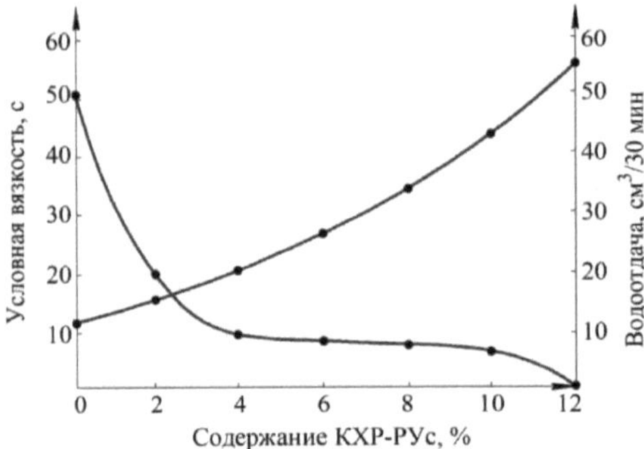

Fig. 4.3. Dependence of conditional viscosity (1) and water yield (2) of the solution on the content of the rapa resistant composite chemical reagent KCD.

Figure 4.3 shows that with increasing concentration of CXR from 2 to 12% the conditional viscosity of the solution increases from 13 to 55 s. The water yield of the solution decreases to 3 cm^3 /30 min with an increase in the concentration of CXR from 2 to 12%.

The increase in viscosity and the noticeable reduction in water yield is, in this case, due to the specific composition and structure of CXR, which allows this reagent to be used as a stabilising base for creating solutions with a high solid phase content.

Further we have carried out laboratory tests on development of formulation of composite chemical reagents for weighted drilling muds using new powdered composite chemical reagents KCR-RUS and different weighting agents in laboratory conditions of SUE "Fan va tarakkiyot". Red clay, barite, hematite and others were used as weighting agents.

The results of studies of physical and chemical properties of weighted drilling muds based on CCHR and various weighting agents are shown in Figures 4.4-4.7.

Figure 4.4 shows the dependence of the density, viscosity and SNA of the drilling fluids on the red clay content.

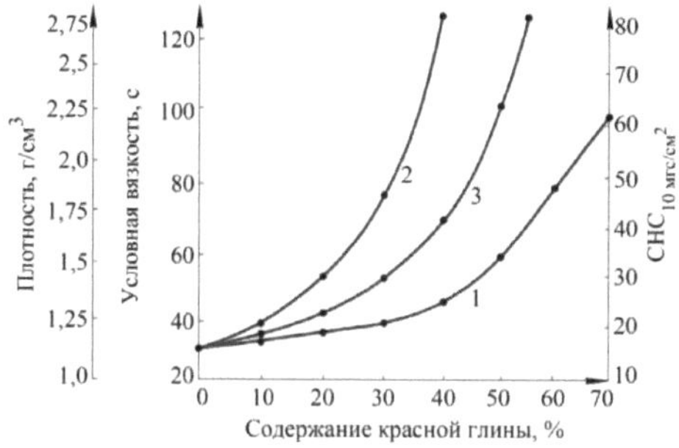

Fig. 4.4. Dependence of density (1), viscosity (2), and SNS_{10} (3), of drilling fluids based on composite chemical reagents on red clay content.

Figure 4.4 shows that as the amount of red clay increases to 40%, the mud density increases to 1.38 g/cm^3 . The mud becomes less fluid with further increase in the amount of red clay, the value of static shear stress is 52 ms/cm^2 . Mud viscosity is within 100-125 sec according to SPV-5 viscometer. The water retention of the solution is 4-5 cm^3 / 30 min and pH 10-11. Thus, the obtained data show that viscosity and static shear stress increase with increasing amount of red clay in the drilling mud. Therefore, for the preparation of weighted muds with densities between 1.17 and 1.38 g/cm^3 red clay can be used in an amount up to a maximum of 40 % of the mud volume.

BaSO sulphate as a weighting agent is widely used to produce super-heavy drilling muds$_4$. Barite weighting agent is produced in Uzbekistan in Tashkent region with density 3,85-3,90 g/cm^3 , in China and Kazakhstan 4,0-4,2 g/cm^3 . In order to develop composite chemical reagents for making weighted drilling muds the studies of physico-chemical properties of weighted drilling muds using barite produced in Uzbekistan were conducted. The results of experimental studies of dependence of physical-chemical properties of drilling muds on barite content are shown in figure 4.5

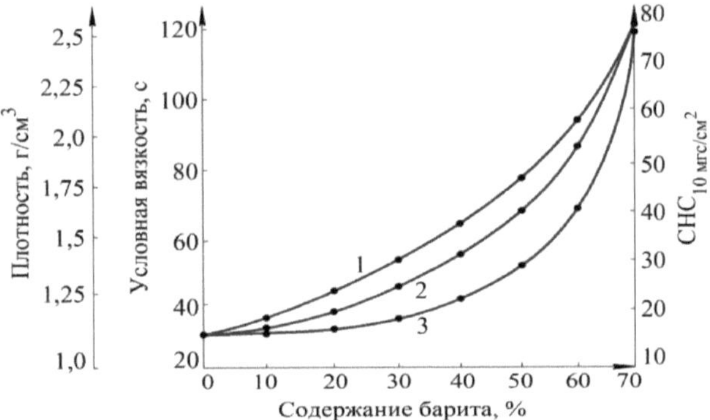

Fig. 4.5 Dependence of density (1), viscosity (2) and SNA_{10} (3) of CCHR-based drilling fluids on barite content

One can see from figure 4.5 that by adding 60-65% of barite weighting agent the mud density is 2.1-2.22 g/cm^3 , while its relative viscosity is 100-112 sec according to SPV-5 viscometer and its SIC is 61-64 mg/cm^2 for 10 min. Mud water yield is 4-5 cm^3 /30 min and hydrogen index is 10-11. Such technological data allow using weighted drilling muds during drilling and opening of salt-and-anhydrite strata of oil and gas wells with abnormally high formation pressure.

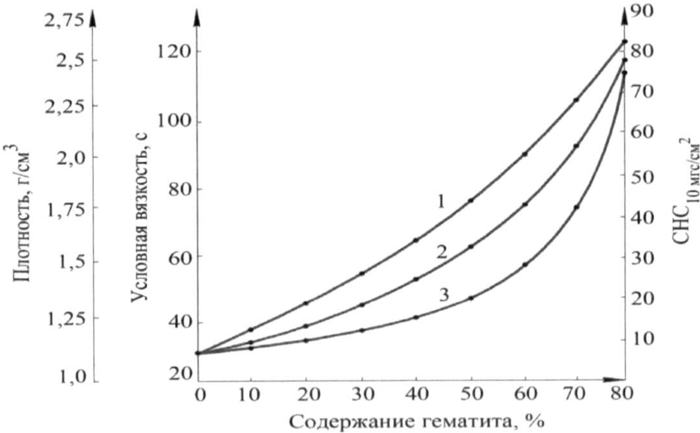

Figure 4.6 Dependence of density (1), viscosity (2) and SNS_{10} (3) of composite chemical drilling fluids on hematite content

In order to obtain extra-heavy drilling muds we carried out experimental studies to study the physico-chemical properties of KCR-based drilling muds using hematite. The results of experimental studies are shown in figure 4.6. One can see from figure 4.6, that if the weighting of drilling mud with hematite is increased up to 70-72%, its density increases up to 2,3-2,65 g/cm^3 and its SNS is within 50-65 mg/cm^2 . The water yield is almost unchanged at 4-5 cm^3 /30 min and a hydrogen index of 11.

Figure 4.7 shows the results of studies of dependence of conditional viscosity of drilling muds based on KCR-RUS and different weighting agents on their density.

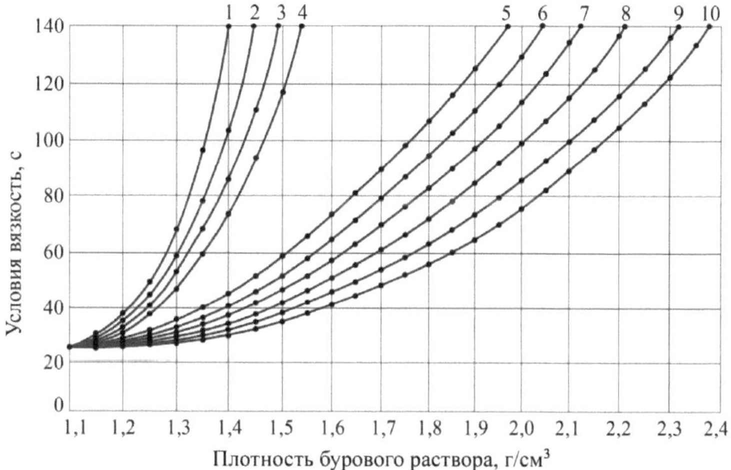

Fig.4.7 Conditional viscosity of drilling muds depending on their density 1-Red clay, 2-KPP, 3-marble, 4-dolomite, 5-barite Uzbekistan, 6-barite Kazakhstan, 7-barite Russia, 8-okalina, 9-magnetite, 10-hematite

As it can be seen from the figure, using red clay in drilling muds prepared on the basis of KCR-RUS one can get drilling mud with density up to 1,4 g/cm^3 , KCSH 1,46 g/cm^3 , marl up to 1.49 g/cm^3 , dolomite up to 1.57 g/cm^3 , Saribulak barite up to 2.03 g/cm^3 , Zheyrem barite up to 2.14 g/cm^3 , Salair barite up to 2.19 g/cm^3 , scale up to 2.24 g/cm^3 , magnetite up to 2.33 g/cm^3 and hematite up to 2.37 g/cm^3 . Further increasing the density of drilling fluids with weighting materials 2.1 g/cm^3 increases the viscosity of the drilling fluids, which leads to an increase in the drag force on the movement of the fluids.

§4.3 Development of technology for production of composite chemical reagents and rapastable drilling muds using the created composite chemical reagents KCR-RUS

Taking into account scientific and methodological principles [28-30] developed by us and identified as a result of comprehensive research, we developed a technological line and created the production of new composite chemical reagents based on local raw materials and waste products at the scientific-production base of scientific and technological centre "NTTS KOMPOZIT". The act of development and creation of technological line is attached (appendix). The following ingredients were used: KPGS, CMC, nedopal, hematite, barite, red clay and bentonite.

Schematic diagram of technological line for production of multiphase rapastable composite chemical reagents developed on the basis of local raw materials and waste products is shown in Fig. 4.8.

As can be seen from the diagram, the technological stage of preparing the components of the barite and nepal weighting compositions and feeding them into the respective aggregates is particularly considered first.

Preparation technology for barite weighting agent. To obtain finely ground barite, first of all, primary granular barite is poured into a container 9 to the full volume. Then, by opening the valve 10, it is fed through the dispenser 11 into the dispenser 25 and undergoes a fine crushing to a particle size of 50-100 microns. The finely ground barite is sieved through sieve 26, goes into the hopper 27 and with the screw feeder 28 is fed into the mixer 33. This is where the powdered chemical reagent is prepared together with the other ingredients.

Powdered polyacrylamide - PAA is poured into container 3 up to its full volume. If necessary, by opening the valve 4, the PAA is fed into the hopper 21 and through the screw feeder 22 into the mixer 33 for the preparation of the composite chemical agent through the batcher 5.

Powdered sodium carboxymethyl cellulose - Na - CMC is poured into tank 6 up to its full volume. If necessary, by opening the valve 7 of the tank 6, Na - CMC enters the hopper 23 through the dispenser 8 and through the screw feeder 24 is fed into the mixer 33 for the preparation of the composite chemical reagent.

Technology of preparation of final products - rapastable composite chemical reagents. Preparation of rapastable composite chemical reagents is carried out as follows: finely ground modified composite powder gossypol resin (CPGS) from screw feeder 18 goes to the paddle mixer 33 for mixing with other components.

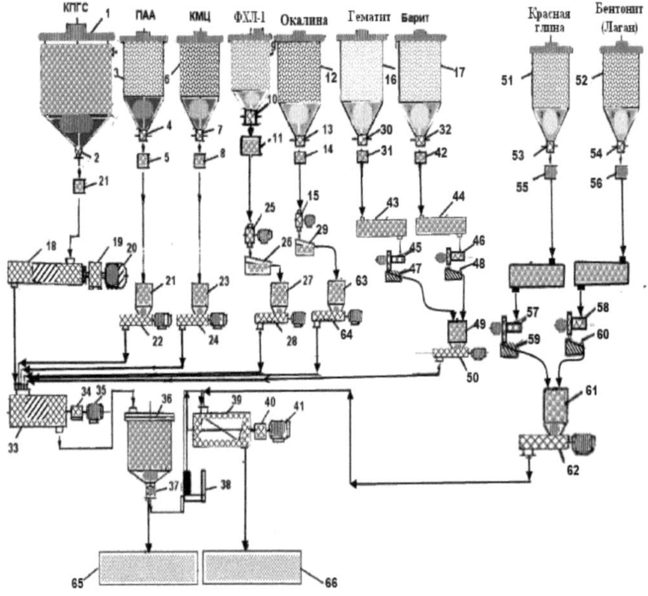

1-Capacitor for CPGS; 2-vent; 3-Capacitor for PAA; 4, 7, 10, 13-flaps; 5, 8, 11, 14, 19, 21-dosing units; 6-Capacitor for KMC; 9- Tank for FHL-1; 12- Tank for scale; 15-dryer; 16- Tank for hematite; 17- Tank for barite; 18, 22, 24, 28, 32; 43-screw feeders; 21, 23, 27, 31, 42-bins; 25, 29-crusher; 26, 30-sieve; 33, 39-mixer; 36-rector for getting of CPR-RU; 38-weights; 43, 44-toppers; 51-topper for clay; 52-topper for bentonite; 53,54-flaps; 55,56-toppers; 57,58-crusher; 59,60-posers; 61-topper; 62-screw feeder; 65,66-packaging lines.

Figure 4.8. Scheme of technological line for production of composite chemical reagents - KCR-RUS for preparation of rapastable drilling muds

In order to obtain the composite chemical KCR-1 10% KMC is added to KPCS. For this purpose PAA or CMC from tanks 3 and 6 through valves 4 and 7 and batchers 5 and 8 are fed to hoppers 21 and 23 and through screw feeders 22 and 24 are fed to the paddle mixer 33. At the same time, barite or nepal is fed through the screw feeders as required. Mixing of the loaded components is carried out within 30-40 minutes until a homogeneous state is obtained. Conditionally called KCR-1Tus reagent goes to the collector 36. At the end of the technological process from the container 36 as a finished product KCR-1TUS is transferred to the packing lines 65 for bagging and storage.

In order to improve physico-chemical properties of drilling muds, 20 to 50 wt.% of nepal, red clay and bentonite are added to KCR-RUS, depending on the need. In these cases the technology of producing the clay reagent is as follows. Underpaste, clay and bentonite from the screw feeder 62, 64 simultaneously with KCG, CMC is fed into the paddle mixer 39 in the same way as KCR-1RUS. The process of mixing of the loaded components is carried out within 30-40 minutes until a homogeneous composite chemical reagent is obtained. Received reagent is called by us conditionally as a composite chemical reagent KCR-2 from the tank 39 in a form of the finished product KCR-2RUS is transferred for packing and warehousing 66.

KCD-1RUS and KCD-2RUS is a ready-made composite material for preparation of drilling muds used in the process of drilling of salt deposits in conditions of rapa-production [36-40].

The abovementioned technological line, as it was mentioned above, is created on the production base of the KOMPOZIT NANOTEXNOLOGIASI Ltd. research and technology centre.

§ 4.4 Conclusions on chapter four

The optimal compositions of chemical reagents of KCD type and properties of drilling muds based on them have been developed and determined. They show the possibility of using weighted drilling mud when drilling oil and gas wells in complicated geological conditions and when penetrating formations with both low and average reservoir pressure. They have surface-active properties. At 0,05 % concentration of drilling agents their surface activity is similar to that of sulfanol and it is considerably higher in comparison to OP-10. Therefore, it is quite possible to use them as a surfactant replacing expensive reagent sulfanol and OP-10.

A resource-saving technology for the production of composite chemical reagents has been developed that makes it possible to produce composite chemical reagents with rational use of raw material resources and modern industrial waste.

CHAPTER V. PRACTICAL AND ECONOMIC ASPECTS OF THE DEVELOPED RAPASTABLE COMPOSITE CHEMICAL REAGENTS AND DRILLING FLUIDS BASED ON THEM

§5.1 Organisation of production of a pilot batch of developed rapa-proof composite chemical reagents used for drilling wells in the conditions of rapa-production

With the purpose of mastering production technology of developed composite chemical reagents based on ingredients from local raw materials and waste products, after fulfilling operations on assembling and start-up and adjustment mentioned above in section 5.1 of this report we carried out works on trial run of technological line in idle mode, making sure of its performance the technological line was launched with KPCS, CMC, PAA, FHL-1, KCR, KPM and bulk ingredients (hematite and barite). Pilot batches of composite rapa-resistant chemicals were obtained for use in well intervention technology in a rapa-production environment.

KOMPOZIT NANOTEXNOLOGIYASI" Ltd. has developed a production line using purchased and available equipment and appliances for producing pilot batches of developed composite flame retardant chemical reagents used in flame-retardant conditions.

In order to continue the researches on studying technological properties of reagents 1000 kg of pilot batches of developed rapastable composite chemical reagents for preparing drilling muds were produced. The laboratory-production research of physical-chemical properties of obtained rapastable chemical reagents KCD-1RUS, KCD-2RUS and KCD-3RUS, physical-chemical and operational properties were carried out; they are presented in chapter 5.

According to the schedule (26.02.2014), the fifth stage of 2015 provides for organization of production and manufacture of a pilot batch of KCD-RUS (KCD-RUS+barite) composite chemicals based on local raw materials with high soluble properties and weighted muds on their basis for drilling of salt deposits in conditions of rapadeposits by KOMPOZIT NANOTEXNOLOGIASI Ltd.

Further on, it is necessary to organise production of a prototype of rapastable weighted drilling muds based on KCD-1-RUS composite chemicals and formation water at one of Uzgeoburneftegaz wells and to test their operational properties.

A pilot batch of 10 tons of KCDR-RUS type efficient rapastatic weighted composite chemicals was **produced at the** technological line established at KOMPOZIT NANOTEXNOLOGIASI Ltd. The act of producing KCDR-RUS type composite chemicals is attached (see attachment).

These rapastable composite chemical reagents with high soluble properties have been submitted for testing in laboratory conditions of Drilling and Cement Mud Service of JSC "Neftegazisproving" JSC "Uzgeoburneftegaz". Table 5.1 shows technological parameters of rapastable drilling mud based on KCR-UR and barite (KCR-RUS) using fresh water.

Table 5.1

Technological parameters of weighted rapastable drilling mud based on KCR-UR (KCR-RUS and barite) and barite (Kirghizia)

№	Composition of weighted mud	Technological parameters				
		ρ, g/sm^3	T_{500} , s	B, sm^3 /30min	K, mm	pH

1	Initial solution 1 litre water+100 g CCR-UR+100 g red clay +100 ml oil+3 g Na_2CO_3	1,09	27	3	3	10
2	Initial solution +500 g barite	1,35	31	3	0,8	10
3	Initial solution +1000 g barite	1,58	35	3	0,8	10
4	Initial solution +1500 g barite	1,72	45	3	0,8	10
5	Initial solution +2000 g barite	1,88	65	3	0,8	10
6	Initial solution +2500 g barite	2,01	110	3	0,8	10
7	Initial solution +2500 g barite +30%NaC1	2,07	105	3,5	0,9	9
8	Initial solution +2800 g barite +30%NaCl	2,21	98	3,5	0,9	9

The results of pilot tests shown in the table above show positive results and fully meet the requirements when using weighted rapastoic composite chemical reagents with high soluble properties in drilling mud for use in drilling wells in the system of JSC "Uzgeoburneftegaz" and are effective.

Further tests were conducted in the laboratory and production conditions of Uzgeoburneftegaz *NHC Uzbekneftegaz*.

§5.2 Pilot testing and use of created rapa-resistant composite chemical reagents while drilling one of the wells of JSC "Uzgeoburneftegaz" under rapa-production conditions

Physico-chemical and technological properties of rapastatic weighted drilling muds based on composite chemical reagent KCR-RUS-1 (KCR-RUS+barite) with high soluble properties have been studied and industrial research on its suitability for producing stable rapastatic clayless and clayey drilling muds from 1,34 to 1,85 g/cm^2 has been conducted.

In the course of work the composition of rapastey weighted drilling mud on the basis of composite chemical reagent (KCR-RUS) at KOMPOZIT NANOTEXNOLOGIYASI LLC and at the laboratory of the Drilling and Cementing Mud Service of JSC "Neftegazisperimental" JSC "Uzgeoburneftegaz" were investigated.

The results of the laboratory tests carried out at KOMPOZIT NANOTEXNOLOGIASI Ltd. are shown in Table 5.2.

Table 5.2

Technological parameters for rapastable weighted of CCR-RUS based clay mud (CCR-UR and barite)

Composition of weighted mud	Technological parameters for weighted mud				
	ρ, g/sm^3	T_{500} , s	B, sm^3 /30min	K, mm	pH
300 ml clay solution +15 g CCR-UR+3 g CarboPAC	1,20	40	6	1	9
300 ml clay solution +15 g CCR-UR +3 g CarboPAC+ 300 gr Barite	1,67	62	6	1,2	9
300 ml clay solution +15 g CCR-UR +3 g CarboPAC+ 400 gr Barite	1,85	80	6	1,3	9

Further, on the basis of results obtained in laboratory conditions of "KOMPOZIT NANOTEXNOLOGIOLOGIASI" Ltd. we carried out laboratory research with the staff of drilling mud and cement slurry service department of "Neftegazis-testing" laboratory of JSC "Uzgeoburneftegaz" on getting formulation of rapastoyl heavy weighted gravelless drilling mud based on composite chemical reagent KCR-RUS. Obtained data are attached in table 5.3.

Table 5.3

Technological parameters of rapastey weighted clayless drilling mud based on KCR-UR and barite (KCR-RUS)

Drilling mud composition	ρ, g/sm^3	T_{500} , s	B, sm^3 /30min	K, mm	pH
1000 ml water + 60 g KHR-RUS +13 g Na-KMC + 3 g PAA +100 ml Oil + 250 g NaCl	1,15	65	4,0	0,1	10
Source solution 1 + 1000 g barite (Uzbekistan)	1,56	142	4,2	0,8	10
No. 2 p-r + 500 g barite (Uzbekistan)	1,74	178	4,5	1,1	10
No. 3 p-r + 200 g barite (Uzbekistan)	1,82	195	4,5	1,1	10

A laboratory test report was approved as a result of the laboratory examination.

From the data of table 5.3 one can see, that technological indexes of weighted clayless drilling mud, processed with composite chemical reagents KCR-RUS+barite have shown their stabilizing advantages of salted weighted clayless drilling mud in all samples, i.e. it reduces filtration properties and allows to get weighted drilling mud density within 1,58-1,82 g/cm^3 .

Further laboratory and production tests were carried out with the composite chemical reagent KCR-RUS (KCR-RUS+barite) to produce a rapastable weighted solution for use in the conditions of rapa manifestations.

Based on the above, it is recommended to carry out further production testing of the KCR-UR composite chemical reagent in the Chulkuvar area of JSC "Uzgeoburneftegaz" fields in the conditions of rapa-evaporation.

The results of tests carried out to produce rapastatic weighted clayey drilling mud based on composite chemical reagent KCR-RUS+barite and working fluid Chulkuvar well #39 are given in table 5.4.

From data of table 5.4 we see, that technological parameters of weighted drilling mud treated with composite chemical reagents KCR-RUS+barite showed all samples their stabilizing advantages of salted weighted drilling mud, i.e. it partially reduces filtration properties and raises pH of the mud.

Table 5.4

Technological parameters for rapa resistant weighted drilling mud based on KCR-RUS and barite

No. n/a	Drilling mud composition	ρ, g/sm^3	T_{500}, s	B, sm^3 /30min	K, mm	pH
1.	1000 ml solution (Chulkuvar 39)	1,34	52	10	1,2	9
2.	No. 1 solution + 100 g NaCl	1,38	64	12	1,5	7
3.	No. 2 p-r +60 g KCR-UR	1,32	69	8	1,0	9

4.	No. 3 Rx +100 ml Oil	1,34	72	8	1,0	9
5	No. 4 p-r +1000 g barite (Uzbekistan)	1,81	195	8	1,0	9
6	After heating at 80^0 C	1,81	90	8	1,0	9
7	No. 4 solution + 5 g Na-KMC , 80^0 C	1,81	148	5,5	0,8	9

A laboratory test report was prepared and approved based on the results of the laboratory examination.

Based on the above-mentioned laboratory and production tests and technological indicators, it can be concluded that new composite chemical reagents are suitable for production of rapaustic drilling muds based on the positive results obtained.

After a laboratory study, the committee recommended that production testing of drilling fluids based on the developed composite chemicals be carried out at well No. 39 Chulkuvar.

The following are the results of the production tests carried out in well Chulkuvar No. 39 and the positive results obtained.

According to the schedule, the seventh stage of the project includes industrial testing of rapastable weighted drilling muds on the basis of composite chemical reagents chemical reagents of KCR-RUS type (KCR-RUS+barite) and their industrial introduction at drilling of oil and gas wells with abnormally high formation pressure in conditions of rapastable events. The act of the conducted tests is enclosed.

Before drilling, we treated an initial working clay solution of 150 m^3 with the KCR-RUS reagent. The initial consumption was 4.8 tonnes of CCR-RUS.

The drilling fluid parameters were as follows: specific whole 1.34, viscosity T-52, water yield B-10, pH-8 and crust 1.2 mm. After treatment with KCR-RUS and other reagents, the mud parameters were as follows: specific whole 1.60, viscosity T-56, water yield B-8, pH-9 and crust 1.0 mm.

The bottom hole depth before our tests was 2,808 metres. The treatment was continued with the original drilling mud with 4.6 tons of CCR-RUS and other reagents (Neft, CMC, Barite). After treatment the parameters of drilling mud were as follows: specific gravity 1.76, viscosity T-58, water yield B-8, pH-9 and crust 0.8 mm.

This solution was drilled to a depth of 2,812 metres.

Further at a face depth of 2812 m. Mud treated with 200 kg of KCR-RUS and CMC-Na (Namangan) and other reagents was used. After treatment, the mud parameters were as follows: specific gravity 1.80, viscosity T-61, water yield B-7, pH-9 and crust 0.7mm. These muds were drilled to 2864 m.

At the bottom hole depth was 2864 m. The initial drilling mud was continued with 200 kg of CXR-RUS and other reagents. After treatment the drilling fluid parameters are as follows: specific gravity 1.81, viscosity T-62, water yield B-6, pH-9 and crust 0.7 mm.

These treated muds drilled down to 2,920m.

The wells went on to reach a depth of 3092 m.

The initial drilling mud treatment with 200 kg of CCR-UR and other reagents was continued. After treatment the drilling fluid parameters are as follows: specific gravity 1.82, viscosity T-63, water yield B-6, pH-9 and crust 0.7 mm. Drilled to 3,200 metres. The results of the production tests carried out are given in Table 5.5.

Table 5.5

Technological parameters of weighted drilling mud based on composite chemical reagent KCR-RUS

Date	Depth-na, m	Composition of treated weighted mud	ρ, g/sm^3	T_{500},s	B, sm^3/30min	pH	K, mm
4.03	2808	Initial clay solution	1,34	52	10	8	1,2

5.03	2808	Initial solution + CHR-Rus (4.8 t) + Neft (4.8 m^3) + CMC-Na (400 kg)+NaCl (3 t)+Barite (82 t)	1,60	56	8	9	1,0
6.03	2808	Initial solution + CHR-Rus (4.6t) + Neft (4.2m^3) + CMC-Na (400kg)+NaCl (3t)+Barite (15.4t)	1,76	58	8	9	0,8
7.03	2808-2812	Initial solution + CHR-RUS (200kg) + Neft (2m^3) + CMC-Na(200kg)+NaCl (3t)+Barite (7t)	1,80	61	7	9	0,7
8.03	2812-2864	Working solution+KHR-RUS(200kg)+Neft(2m^3) +KMC-Na(200) kg)+Barite (8t)	1,81	62	6	9	0,7
9.03	2864-2920	Working solution+KHR-RUS(200kg)+Neft(2m^3) +KMC-Na(200) kg)+Barite (9t)	1,82	63	6	9	0,6
10.03	2920-3092	Working liquor + Calcium soda (320g) + Carbo PAC (500kg) + Caustic soda (100kg) + Barite (8t)	1,83	65	5	9	0,5
11.03	3092-3200	Working liquor+Calcified Soda (400kg)+Carbopack (500kg)+Caustic Soda (100kg) +Barite (6t)	1,84	68	5	9	0,5

Table 5.5 shows the compositions of prepared weighted mud and its technological parameters, as well as daily meters of penetration. After drilling to the depth of 3200 m. jointly with specialists of JSC "KashPI" JSC "Uzgeoburneftegaz" prepared an act of introduction (act of introduction is attached in the appendix) with chemical reagents KCR-UR, formulation of weighted drilling mud and its efficiency, and also this composition is recommended for drilling well from salt and anhydrite formation in the fields of JSC "Uzgeoburneftegaz".

Therefore, based on pilot testing and positive results, we believe that new chemical reagents KCD-1RUS and KCD-2RUS are suitable for treatment of drilling muds based on highly saline formation water (during drilling under rape conditions) and it is recommended to expand its use in wells drilling within JSC "Uzgeoburneftegaz".

Thus, the technology of producing KCR-RUS composite chemical reagents and drilling muds based on them has been mastered.

Later on, in order to introduce them on a large scale in the industry and for the purpose of recommendation we have developed normative and technical documents: technological regulations for obtaining KCR-RUS composite chemical reagents and drilling muds on their basis, technical conditions for them and the standard of the companies.

§5.3 Calculating the technical and economic efficiency of using composite chemical reagents and Rapastreamer-based drilling fluids

For production of clayey weighted drilling mud used in the process of drilling oil and gas wells, as noted above, mainly barite, hematite, K-4, PAA, CMC, PCLS, oil, caustic soda and soda ash are used, the quantity and contract price of which are shown in Table 5.6.

During the pilot test we used a new formulation of drilling mud using the composite chemical reagent KCR-RUS, oil; barite and hematite from the Bekabad field to replace the existing clay mud (Table 5.6).

Table 5.6

Quantity and cost of clay, chemical and composite reagents for preparing 1 m^3 of drilling mud

Reagent name	Weight and cost of clay and reagents for existing drilling mud		Weight and cost of clay and reagents for drilling mud using CIIR-RUS		Cost of 1 tonne of clay and reagents,
	ˣWeight,kg	Price,amo	ˣWeight,kg	Price,amou	
Barite or hematite	250	31250	250	31250	125000
K-4	100	90000	50	45000	900000
CMC	50	250000	-		5000000
Caustic soda	60	81400	-		1356667
Calcined soda	5	3750	-		750000
PAA	40	86600	-		2165000
CHR-RUS	-	-	100	75000	750000
Oil	100	20000	100	20000	200000
Total amount	605	563000	500	171250	

Notes:ˣ the flow rate of mineralised, water in preparation of 1 m^3 of drilling mud is 700 litres.

It follows from the above that when replacing 1m^3 of drilling mud currently used in drilling oil and gas wells in the Chilkuvar field with 1m^3 of mud based on the KCR-Rus we have developed, the economic efficiency will be as follows

$$\textbf{C} = \textbf{Price}_1 - \textbf{Price}_2 = 563000-171250 = 391750 \text{ sum}$$

where: Ts_1 is the price of 1m^3 of the existing drilling mud currently in use;

Ts_2 - price of 1m^3 of the proposed drilling mud using the developed composite chemical reagent KCR-RUS.

During pilot testing at the well №39 Chilkuvar JSC "KashPI" used only 210 m^3 of drilling mud using new composite chemical reagent KCR-UR. At the same time the economic efficiency of 210 m^3 of drilling mud with the use of KCR-UR during drilling was as follows

$$\textbf{E} = \textbf{V}_1 \textbf{ x} \textbf{E}_1 = 210 \text{ m}^3 \text{ x } 391750 = \text{SUM } 82{,}267{,}500$$

where $\textbf{V}_1$ is the volume (210 m^3)of drilling fluid based on KCR-RUS for oil and gas drilling.

$\textbf{E}_1$ - economic efficiency of applying 1 m^3 of drilling mud using KCR-RUS.

The actual economic efficiency of applying 10 tonnes of CXR-Rus in drilling well No. 39 Chilkuvar of KashPI JSC amounted to 81.95 million soums.

Using 1,000 tonnes of composite chemicals and salt-resistant drilling mud on their basis when drilling wells would have an expected economic effect of about SUM 8.2 billion.

§ 5.4. Conclusions of chapter five

Extended testing and experimental implementation of experimental batches of KCD-RUS, drilling fluids based on them in laboratory and industrial conditions during drilling of oil and gas well No.39 Chulkuvar with participation of engineering and technical personnel of JSC "KashPI"

JSC "Uzgeoburneftegaz" were carried out. Positive results have been obtained, which are reflected in the acts of relevant tests.

Technological regulations for production of pilot batches of composite chemical reagents of KCD type and technical conditions for drilling muds produced on their basis were developed.

Actual technical and economic efficiency of implementation of developed composite chemical reagent compositions of KCD-RUS type when using them in drilling mud during drilling of oil and gas well #39 at Chulkuvar field of JSC "KashPI" AK "Uzgeoburneftegaz" in 210 m.[3] amounted to 81.95 million soums. When using 1000 tons of composite chemical reagent in well drilling the expected economic effect will be more than 8.2 billion soums.

CONCLUSION

1. For the first time the possibility of creating rapastable compositions of chemical reagents for drilling fluids and the technology of their production on the basis of local raw materials and industrial wastes was scientifically grounded.

2. Main regularities of influence of organic-mineral ingredients of composite chemical reagents, leading to the improvement of physical-chemical, technological and operational characteristics of rapastable drilling muds, allowing to drill oil-and-gas wells in different depths of saline formations, have been determined.

3. For the first time optimal compositions of chemical reagents of KCR-RU class for preparation of rapastable drilling mud have been developed and established that they show surface-active properties and are practically not inferior to parameters of drilling mud prepared by commonly used reagents.

4. It is established, that the introduction of the developed composite chemical reagents of KCR-RUS class into the composition of drilling mud, promotes the manifestation of synergistic effect and improvement of viscosity-rheological properties, which provide stabilizing properties of drilling fluids.

5. It has been revealed that drilling mud prepared on the basis of developed composite chemical reagents of KCR-RUS class accelerates mechanical drilling speed by 10-15%, increases penetration of productive oil and gas horizons by 30-35%, and also improves environmental safety.

6. The organisation's standard (technical specifications) and technological regulations for production of rapastable composite chemical reagents based on organomineral ingredients from local raw materials and industrial waste have been developed.

7. The application of created new composite chemical reagents of KCR-RU class for preparation of drilling muds not only in fresh, but also in highly saline formation water for oil and gas wells in the system of JSC "Uzbekneftegaz" is recommended.

REFERENCE LIST

1. Sh.M. Mirziyoyev. We shall build a free, democratic and prosperous State in Uzbekistan together with our courageous and noble people. - T.: Narodnoe slovo N 248(6653). 16.12 2016 г
2. Likhushin A.M., Lavrentyev V.S., Nifantov V.I. et al. Well driving in complicated mining and geological conditions //Gas Industry, 1998. - № 10. - C. 40-42.
3. Lukyanov E.E., Strelchenko V.V. Geological and technological studies during drilling. - Moscow: Oil and Gas, 1997. - 688 c.
4. Panteleev A.S., Kozlov N.F. Geological structure and oil and gas bearing capacity of the Orenburg region. - Orenburg: Orenburg Book Publishers, 1997. - 272 c.
5. Abaturov V.G. Drilling in complex geological conditions. Part 1. Accidents, their prevention and elimination: a course of lectures. - Tyumen: TyumSOGU, 1995.- 60 p.
6. Bulatov A.I., Proselkov Y.M., Shamanov S.A. Technique and technology of oil and gas wells drilling: Textbook for Universities. - Moscow: OOO Nedra-Business Center, 2003g. - 107 c.
7. Basarygin Y.M., Bulatov A.I., Proselkov Y.M. Drilling of oil and gas wells. - M: Nedra-business-center, 2002. - 632 c.
8. Shamanov S.A. Drilling and injection of horizontal wells. - M: Nedra, 2001. - 190 c.
9. Ryazanov Y.A. Encyclopaedia on drilling fluids. - Orenburg:Chronicle, 2005. - 664 c.
10. Drilling Fluids Handbook //Drilling, completionandworkoverfluids. Supplement to the journal "Oil and Gas Technologies". -M:Fuel and Energy, 2007. - 48 c.
11. Gorodnov V.D. Drilling muds. - Moscow: Nedra, 1985. - 206 c.

12. Ananyev A.N. Training Manual for Drilling Fluids Engineers. - Volgograd: International Casp Fluids, 2000. - 139 c.
13. Kister E.G. Chemical Treatment of Drilling Muds, Moscow, Nedra, 1972 - 392 pp.
14. Basarygin Y.M., Bulatov A.I., Proselkov Y.M. Complications and accidents during drilling of oil and gas wells. -M.: OOO "Nedra-Business Centre", 2000.p - 679.
15. Basarygin Y.M., Budnikov V.F., Bulatov A.I. Technological bases of development and killing of oil and gas wells. -M: Nedra, 2001.p-543
16. Shandin S.N., Ryabokon S.A., Shevkina Z.A. Gravity barite weighting agent. "Drilling", 1976, No.5, pp. 13-15
17. Ryabokon S.A., Penkov A.I. Prevention of flocculation of barite weighting agents in oil emulsion solutions.-"Drilling" 1974, No. 6 p. 17-20
18. Minhairov K.L. Naumov V.G. Petroleum-based heat resistant weighted mortar. 1970., 170-176 c
19. Ryabchenko V.I. Management of drilling mud properties. -M: Nedra, 1990. - 230 c.
20. Gorodnov V.D. Drilling muds. - Moscow: Nedra, 1985. - 206 c.
21. Bulatov A.I. Fundamentals of physical chemistry of washing liquids and plugging solutions. -M.:Nedra, 1968. - 424 c.
22. Bulatov A.I. Theory and practice of well injection. -M: Nedra, 1998. - 496 c.
23. Bulatov A.I., Makarenko P.P., Proselkov Y.M. Drilling Washing and Plugging Solutions - M: Nedra, 1999. - 424 c.
24. Paus K.F. Drilling Muds. -M.:Nedra, 1967.
25. Paus K.F. Drilling fluids. -M.:Nedra, 1967.

26. Flushing Solutions for Drilling Wells. Collection. -M.: Gostoptekhnizdat, 1962.
27. Preparation and application of flushing and plugging dispersions in drilling. Collection. - Kiev: Naukova Dumka, 1984.
28. Miskarli A.K., Bayramov A.M. New surfactants for oil drilling. - Baku, 1964. - 24 c.
29. Lityaeva Z.A. Clay powders for drilling fluids. -M: Nedra, 1992. - 192 c.
30. Korotkova E.I., Gindullina T.M. Physico-chemical methods of research and analysis. Publishing house of Tomsk Polytechnic University, 2011. - 168c.
31. Gorodnov V.D., Rusaev A.A. The role of composition of cation-exchange complex of clays in their stability. Dispersed Systems in Drilling. Kiev: Naukova Dumka, 1977, pp. 91-93.
32. Ghamzatov S.M., Rakhimbaev Sh.M., Rakhmatov R.M. et al. Influence of clay genesis on behaviour of clay sediments during drilling and well support. Express Information: Geology, drilling and development of gas fields. No.13 1976, pp. 3-4.
33. Vafin R.M. Development of compositions of weighted drilling muds based on cuttings for drilling and completion of wells with abnormally high formation pressures. D. thesis of Candidate of Technical Sciences-SanktPeterburg. Sanct-Petersburg State University of Oil and Gas-2012, -133 pp.
34. Akhmedov R.G. Features of drilling wells in clayey sediments. Moscow VINITI, 1977, p. 179.
35. Gorodnov V.D. Physical and chemical methods of preventing complications in drilling. Moscow. "Nedra", 1977, 280 p.

36. Zubarev V. G. G. Dynamics of cavern formation process. Drilling magazine. No 4, 1970, p. 37-39.
37. Zubarev V.G., Baidyuk B.V. Study of penetration of leachate of washing fluids into clayey rocks. Express Information. VNIIEGazprom, No.4, 1973, p44.
38. Avetisyan N.G., Shemetov V.Yu. Prevention of rock stability disturbances under osmotic mass transfer. Review information. VNIIOENG, Series: Drilling. 1980г.
39. Peslyak Y.A. Behaviour of clays during drilling and well operation. Journal "Neftyanoye bukhodya". No.11, 1960, pp. 28-34.
40. Gusakov N.A., Priverdyan A.M. On determining hydrodynamic pressure on well walls during lowering and lifting operations. Moscow, "Neftyanoye upravlenie", No 8, 1959.
41. Lopatin V. A. A., Mukhin L. K. Influence of hydraulic pressure on stability of clayey rocks during well drilling. Izvestiya Vuzov. Series: Oil and Gas, No. 6, 1964.
42. Abramzon. A.A. Surfactants. / Abramzon. A.A., Bocharov V.V. et al. - Moscow: Chemistry, 1979, 376 pp.
43. Bulatov A.I. Reference book on well flushing / M.: Nedra. 1984. 317 c.
44. Ryazanov Y.A. Encyclopedia on drilling fluids / Nedra. 2004, 490 c.
45. Akhmadeev R.G. Chemistry of flushing and plugging fluids / M.: Nedra. 1981. 152 c.
46. Lityaeva Z.A. Clay powders for drilling fluids / M.: Nedra. 1999. 192 c.
47. Gray J.R. Composition and Properties of Drilling Muds (Flushing Fluids) / M.: Nedra. 1985. 159 c.
48. Gorodnov V.D. Drilling Muds / M.: Nedra. 1985. 206 c.

49. Mavlyutov M.R. Technology of deep well drilling / M: Nedra. 1982. 287 c.

50. Ibragimov G.Z., Fazlutdinov K.S., Khisamutdinov N.I. Application of chemical reagents for oil production intensification/Moscow: Nedra. 1991. 356-382 c.

51. Dedusenko G.Ya. drilling fluids with low content of solid phase / M.: Nedra. 1985. 159 c.

52. Yarov A. N. Drilling muds with improved lubricating properties. M.. Nedra, 1975,143 p.

53. Mezhlumov A.O. Complications and accidents during well drilling with the use of gaseous agents. - Moscow: Nedra, 1970, 193 p.

54. Sherstnev N. M. M. Prevention and elimination of complications in drilling. - M.: Nedra, 1977, 304 p.

55. Edgorov N. Chemical reagents and materials in the oil and gas sector. LLC and Voris-Nashrioti. Tashkent, 2009, 519 pp.

56. Paus K.F. Drilling Fluids.-M.: Nedra, 1973, 303 p.

57. Ananyev A.N. et al. Drilling Muds for Well Conducting in Complex Conditions of Lower Volga Region. RNTS "Drilling", No.10, 1976.

58. Angelopulo O. K., Podgornov V.M., Avakov V.E. Drilling muds for complicated conditions. -M.: Nedra, 1988. -135 c.

59. Gorodnov V.D. Drilling muds. - Moscow: Nedra, 1985, 206 p.

60. Bulatov A.I. et al. Drilling muds. - Moscow: Nedra, 1999. 425c.

61. Ryazanov Y. A. A. Reference Book on Drilling Muds. - Moscow: Nedra, 1979, 215 p.

62. Ryazanov Y. A. Encyclopedia of drilling fluids. -M.:Nedra, 2004, 490 p.

63. Degtev N, I. Control and degassing of drilling flushing fluids. - Moscow: Nedra, 1978, 152 p.

64. Gray J.R. Composition and properties of drilling agents (flushing fluids). - Moscow: Nedra, 1985, 509 p.
65. Ryabchenko V.S. Voytenko Managing rock pressure while drilling wells.- M.: Nedra, 1985, 181 p.
66. Voytenko V.S. Combatting manifestations of rock pressure in deep wells.
67. Vinnichenko V. M. Prevention and elimination of accidents and complications during drilling of exploration wells. - Moscow: Nedra, 1991, 170 p.
68. Bochko E. A. Strengthening of unstable rocks during borehole drilling. - Moscow: Nedra, 1979, 168 p.
69. Bulatov A.I. Drilling of oil and gas wells. -M.:Nedra, 2001,496 p.
70. Badovsky N.A. Combat complications during drilling deep wells abroad.- M.: VIEMS, 1986, 57 p.
71. Rabia H. Technology of drilling wells. - Moscow: Nedra, 1989,270 p.
72. Shmelev P. S. Drilling of deep wells in conditions of anomalous impact of corrosive-active media. - Moscow: Nauka. 1998, 351 c.
73. Shevtsov V.D. Combat with blowouts during well drilling. - Moscow: Nedra, 1977, 133 pp.
74. Surikova O. A. A. Prevention and elimination of complications in fractured rocks. - Moscow: VNIIONG, 1985, 58 p.
75. Yasov V.G. Complications in drilling. - Moscow: Nedra, 1991, 334 p.
76. Mamajanov U.D. Dynamic characteristics of flushing solutions and complications in drilling. - Leningrad: Nedra, 1972. - 163c.
77. Rakhimov A.K. Stripping and anchoring of wells in conditions of abnormally high reservoir pressures. - Tashkent: Fan, 1980. - 170 c.
78. Yagudin S.Z. Brine fountains from salt deposits and dealing with them. // NTS "Drilling". Moscow, 1968. - №10.- c. 18-20.

79. Aminov A.R. Drilling of deep wells in complicated conditions. Tashkent, Fan, 1992.

80. Adamson A. Physical chemistry of surfaces / Edited by 3. M. Zorin and V.M. Muller.-M.: Mir, 1979, 568 p.

81. Voyutsky S.S. Course of colloidal chemistry. - Moscow: Chemistry, 1976, 512 p.

82. Gorodnov V.D. Physico-chemical methods of preventing complications in drilling. - Moscow: Nedra, 1977, 280 p.

83. Kruglyakov P. M., Rovin Yu. G. Physico-Chemistry of Black Hydrocarbon Tribes nok. - Moscow: Nauka, 1978, 183 pp.

84. NeumanA. W., GoodR. J.- In: Surface and Colloid Science, v. 11 / Ed. R. J.

85. Good and R. R. Stromberg. New York. Plenum Press, 1979, p. 31-92.

86. Hansen R. S. - J. Appl. Phys., 1964, v. 35, № 6, p. 1983.

87. Mann J. A., Hansen R. S. - J. Coll. Sci., 1963, v. 18, p. 757.

88. *Mapp* J. A., Hansen R.S. - J. Appl. Phys., 1964, v. 35, № 1, p. 152-158.

89. Paddey J. F. Surfase and Colloid Science, v. 1, 1964, p. 9-251.

90. Abramzon A. A. Surfactants: properties and applications. - L.: Chemistry, 1975.

91. Abramzon A. A., Bocharov V. V., Chaevoy G. M. et al. Surface-active substances. - Moscow: Khimiya, 1979, 376 p.

92. Dedusenko G.Y. Drilling muds with low content of solid phase. Moscow: Nedra, 1985,159 p.

93. Pugachevich P. P., Beglarov E. M., Lavygin I. A. Surface phenomena in polymers. - M.: Chemistry, 1982, 200 p.

94. Lipatov U. S., Nesterov A. E., Gritsenko T. I. Reference book on polymer chemistry. - Kiev: Naukova Dumka, 1971, 536 p.

95. Rytslin R.E. Stabilisation of clay solutions on Zhetybai field // NTS "Drilling", 1964. - 12. - c. 11-13.
96. Pozin M. E. Technology of mineral salts. In 2 vols. - M.: Chemistry, 1970.
97. Zhukhovitskiy S.Y. Flushing liquids in drilling. -M.:Nedra, 1976, 200 p.
98. Emulsions. Edited by F. Sherman/Translated from English: Chemistry, 1972, 448 p.
99. Vydra V. A. Corrosion and protection of fasteners of oilfield equipment.
100. Akhmetshin E. A. Drilling wells in hydrogen sulphide conditions. - Moscow: VNII IONG, 1983, 43 p.
101. Gutman E.M. Protection of oilfield equipment from corrosion. - Moscow: Nedra, 1983, 152 p.
102. Patent RU 2385892S09K8/20 A method of stabilisation of the salt-saturated drilling mud. Bruy Lubov Kozminichna, Tolkacheva Tatiana Maksimovna, Prizentsov Alexander Ivanovich, Tereshchuk Lubov Ivanovna, Dobrodeyeva Inna Vladimirovna, Paskaru Konstantin Grigorievich10.04.2010
103. Negmatova K.S., Salimsakov Y.A., Rakhimov H.Y., Sharifov G.N., Kobilov N.S. Use of wastes of oil and fat production and getting of effective composite oil emulsion drilling fluids // Kimyonin dolzarb muammolari RIAC materialallari. (IIkism). - Samarkand, 2009.-P.65-66.
104. Negmatova K.S., Negmatov S.S., Salimsakov Y.A., Rakhimov H.Y., Negmatov J.N., Isakov S.S., Sharifov G.N., M.I. Negmatova. Research Of The Structure And Properties Of Viscous Gossypol Resin Powder And The Mechanism Of Formating Of Water-Soluble

Chemical Composition OnIts Base. // 6th international conference on times of polymers (top) & Composites prof. Alberto d'amore . Italy. 2012.

105. K. Negmatova, Sh. Isakov, N. Kobilov, M. Negmatova, J. Negmatov, J. Haydarov, G. Sharifov, Sh. Rahimov. Effective Composite Chemical Reagents Based On Organic And Inorganic Ingredients For Drilling Fluids Used In The Process Of Drilling Oil Wells. // Advanced Materials Research Vol. 413 (2012). Trans Tech Publications, Switzerland. P. 524-527.

106. Negmatova K.S., Rakhimov H.Y., Sobirov B.B., Rakhmonov B.Sh., Negmatov S.S., Salimsakov Y.A. Technology of powder gossypol resin multifunctional purpose // Academic Journal of Western Siberia. Natural Sciences: Achievements of the new century, 2011. - № 2. - C.64-65.

107. Sharifov G.N., Kobilov N.S., Negmatova K.S., Rakhimov H.Y. Thermal resistant chemical reagents for drilling fluids and their classification \\International Scientific and Technical Conference "Resource and energy saving, environmentally friendly composite materials". 19-21 September - Tashkent, 2013. - C. 323-325

108. Rakhimov K.Y., Negmatova K.S., Kobilov N.S., Negmatov S.S., Dusmuratov E.B., Sharifov G.N., Berdiyev S., Mamanov B., Razhabov A., Raupova D. Study of influence of optimal composition of composite emulsifier KP-GEM on technological parameters of drilling fluids// Materials of RSTC "Progressive technologies of composite materials and products from them". - Tashkent, 2015. - C. 235-236.

109. Negmatova K.S. Technology of composite chemical reagents and drilling fluids and their efficiency. - Tashkent: SUE "Fan va tarakkiyot", 2015. - 48 c.
110. Negmatova K.S. Development of effective compositions of composite chemical reagents and lightweight drilling fluids based on them// Composite materials. - Tashkent, 2015. - №1. - C. 4-7.
111. Negmatov S.S. Progressive technologies of reception of composite materials and products from them. //Materials of RSTC "Progressive technologies of obtaining composite mtaerials and products from them". -Tashkent, 2015. -C.3-5.
112. Sharifov G.N., Negmatova K.S., Kobilov N.S., Negmatov J.N., Rakhimov H.Y. Composite chemical reagent for drilling mud of saline-anhydrite formation in rapa-propagation. Composite Materials Journal. Toshkent, 2016, №3, -P.116-118.
113. Negmatov S.S., Negmatov J.N., Negmatova K.S. Kobilov N.S., Sharifov G.N. Chemical reagents for stabilization of non-tightened and tightened drilling muds allowing the release of mobility loss of drill string and casing during drilling of oil and gas wells/ Jour. "Composite materials", - Tashkent, 2015, - №3. - C. 85.
114. Kobilov N.S., Negmatova K.S., Sharifov G.N., Negmatov S.S. Chemical reagent KCR-UR for production of weighted clayless drilling muds Materials of RSTC "Progressive technologies of composite materials and products from them". - Tashkent, 2015. - C. 237-238.
115. Rakhimov H.Y., Negmatova K.S., Kobilov N.S., Negmatov S.S., Dusmuradov E.B., Sharifov G.N. Study of influence of optimal composition of comopositional emulsifier KP-GEM on technological parameters of drilling fluids. Materials of RSTC "Progressive

technologies of composite materials and products from them". - Tashkent, 2015. - C. 239-240

116. Kobilov N.S., Rakhimov H.Y., Negmatova K.S., Sharifov G.N., Abdukarimov M, Raupova D.N. Study of composite polymer chemical reagent effect on physicochemical properties of weighted drilling mud. // Materials of Scientific and Technical Conference "The role of integration of polymer science and education in the innovative development of industries" Tashkent, Research Centre of Chemistry and Physics of Polymers at National University of Uzbekistan. 2015 in press.

117. Sharifov G.N., Negmatova K.S., Kobilov N.S., Rakhimov H.Y., Raupova D.N. Composite chemical reagent for drilling fluids during drilling of salt-and-anhydrite strata in rapaproduction. // Materials of Scientific and Technical Conference "The role of integration of polymer science and education in the innovative development of industries" Research Centre of Chemistry and Physics of Polymers at the National University of Uzbekistan. Tashkent, 2015.

118. Sharifov G.N., Negmatova K.S., Kobilov N.S., Rakhimov H.Y. Development of rapastable composite chemical reagents based on organic and inorganic ingredients for drilling fluids used in oil and gas wells drilling. Journal. "Composite materials", - Tashkent, 2015, - №4. - C. 124-125.

119. Sharifov G.N. Negmatov S.S. Development of effective composite chemical reagents for drilling fluids used in drilling of oil and gas wells with rasto-impregnation // Proceedings of the RSTC "Ingredients from local and secondary raw materials to produce new composite materials", -Tashkent, 2014. -C. 284-286.

120. Kobilov N.S., Negmatova K.S., Rakhimov H.Y., Sharifov G.N., Abdukarimov M., Raupova D.N. Production testing of composite

chemical reagents for weighted drilling muds during drilling of salt and anhydrite strata \ Prospects of composite and nanocomposite materials: Materials of Republican Scientific and Technical Conference - Tashkent, 11-12 November 2016. - C.124-126.

121. Sharifov G.N., Negmatova K.S., Kobilov N.S., Rakhimov H.Y.,Negmatov J.N. Development of effective thermo-salt-resistant composite chemical reagents for drilling fluid\composite materials. - Tashkent, 2017. - №1, -C. 94-95.

122. Sharifov G.N. Development of effective composite chemical reagents for salt-resistant clayless drilling mud. Materials of republican scientific-technical conference "Prospects of development of composite and nanocomposite materials". April 5-6, 2018.

123. Negmatov S.S., Sharifov G.N., Negmatova K.S., Kobilov N.S., Rakhimov H.Y. Composite polymeric material is multifunctional effective chemical reagent for drilling fluids\ Materials of Republican scientific and technical conference "New composite and nanocomposite materials: structure, properties and application", April 5-6, 2018. -Tashkent, - P. 187-188.

124. Sharifov G.N. Development of effective composite chemical reagents for drilling mud in salt-and-anhydrite strata\\"Ўzbekistonning қtisodiy rivozhlanishida kimyoning urni" mausudagi republic ilmiy-amali anjumani materialari 24-25 May - Samarkand, 2018.-P.45-46.

125. Sharifov G.N. K.S. Negmatova, N.S. Kobilov, S.S. Negmatov, B.S. Egamberdiev, Y.K. Rakhimov New compositions of weighted drilling muds for oil and gas well drilling\ Innovative developments in chemistry and technology of fuels and lubricants: Collection of reports and abstracts of III International Scientific and Technical Conference. ONC Institute, 19-20 September - Tashkent, 2019. - C. 191-192.

126. Sharifov G. N. The current state of chemical reagents used in oil and gas drilling wells\ Innovative developments in the field of chemistry and technology of fuels and lubricants: Collection of reports and abstracts of III International Scientific-Technical Conference. ONC Institute, 19-20 September - Tashkent, 2019. - C. 193-194.

127. Sharifov G.N. K.K. Rakhimov, S.S. Negmatov, M.T. Anvarova, H.Y. Rakhimov. Determination of chloride salt content in oil from Sovligar field by indicator titration of water extract\\ Resource- and energy-saving, environmentally friendly composite nanocomposite materials. Republican scientific and technical conference. April 25-26, Tashkent, 2019. - C. 210-211.

128. Sharifov G.N. Kobilov N.S., Negmatova K.S., Negmatov S.S. Organomineral weighting ingredients for producing weighted drilling muds used in oil and gas well drilling\ Resource- and energy-saving, environmentally friendly composite nanocomposite materials. Republican scientific and technical conference. April 25-26, Tashkent, 2019. - C. 195-199.

129. Negmatova K.S., Anvarova M.T., Sharifov G.N., Ergasheva S.H., Kadirov H.I., Turabjanov S.M. Composition of corrosion inhibitors for saline acid treatment of wells\ Composite materials 1/2019 P.15-17

130. Sharifov G.N. Study of physical-chemical and technological properties of drilling muds using newly developed composite chemical reagents \ Central Eurasian Studies Society International Scientific-online Conference on innovation in the modern education system. Washington. 2021.C. -335-342.

Printed by Books on Demand GmbH, Norderstedt / Germany